THE LOOMING ENERGY CRISIS

Other Books by the Author

Energy: The Source of Prosperity

Crisis in the Mideast (a novel)

Nothing to Fear

CLEXIT

Carbon Folly

Entrepreneur as CEO

THE LOOMING ENERGY CRISIS

ARE BLACKOUTS INEVITABLE?

DONN DEARS

SEA·HILL
PRESS

Dears, Donn D.

Looming Energy Crisis: Are Blackouts Inevitable?

Includes Index

ISBN 978-0-9815119-9-3

Published by Seahill Press.

Manufactured in the United States.

Dedicated to my grandchildren

Acknowledgments

It's not possible to write a book as complex as this without help from others.

Elliott Seiden, on reading the original draft, provided valuable insights that I had overlooked. I use his words when I describe the importance of preparing for the next pandemic by ensuring the availability of reliable electricity. I restructured the book based on his comments.

His experience as Section Chief in the Antitrust Division of the Department of Justice, and then as a lawyer in the airline industry, helped me understand how the Federal Energy Regulatory Commission (FERC) was overstepping its authority. While I thought FERC was acting inappropriately, I was not aware there was a legal term, arrogate, to describe such actions.

Meredith Angwin has been counseling me on the actions of the Independent System Operator New England (ISO-NE) for the past few years. She has spent years trying to understand the inner workings of the Regional Transmission Organizations and Independent System Operators (RTO/ISOs), including four years as a member of the Coordinating Committee of the Consumer Liaison Group of ISO-NE.

She is writing a new book, *Shorting the Grid, The Hidden Fragility of our Electric Grid*. I'm hopeful she can get it published quickly. While I haven't yet read it, I'm sure it will provide much needed insight into the functioning of RTO/ISOs.

Meredith reviewed some of my writing to ensure I had fully understood how certain procedures work. In addition, Meredith has also been a tireless campaigner for nuclear power.

Sue Ann Baker, author of *Behind the Shades: a Female Secret Service Agent's True Story*, reviewed the text to help me avoid technical jargon and ensure that technical issues were clear.

Dr. William Happer reviewed my descriptions of the curves he had published establishing how little temperatures would rise with a doubling of carbon dioxide and of methane.

Dr. Happer is a world-renowned physicist and is the Cyrus Fogg Brackett Professor of Physics Emeritus at Princeton University.

It has been my purpose throughout to be certain that all the facts in the book are accurate.

I hasten to add that each of these individuals do not necessarily agree with any of the conclusions I have reached in this book.

DONN DEARS
MAY 2020

Contents

Illustrations

Figures

Tables

The greatest challenge facing mankind is the challenge of distinguishing reality from fantasy, truth from propaganda. So it's time to abandon the religion of environmentalism, and return to the science of environmentalism, and base our public policy decisions firmly on that.[1]

Michael Chrichton

Commonwealth Club, San Francisco, CA

September 15, 2003

Introduction

When we flip the switch, the lights come on without anyone thinking about it. This has only been true for the last hundred years in metropolitan areas, and for only approximately eighty years in rural areas with the enactment of the Rural Electrification Act of 1936.

In 1935, only 25 percent of rural homes in the United States had electricity,[2] and there are people alive today who grew up without electricity. Today, few people are even aware of the monolithic system that generates, distributes, and controls the electricity that flows with seeming effortlessness across the United States. This system is referred to as the grid, which is actually three grids covering the entire lower forty-eight states.

Over the past one hundred years, there have been only two area-wide blackouts affecting over 30 million people caused by a failure of the transmission system. There have been other blackouts—mostly caused by storms—affecting smaller groups, perhaps as many as several million people. Overall, the grid has worked remarkably well. Reliability can still be improved upon, but this is primarily a question of placing transmission and distribution lines underground to minimize weather-induced outages.

Suddenly, we are faced with a threat to the grid we haven't seen before. It is a threat that can dramatically increase blackouts and the suffering that accompanies them. Some in leadership positions have viewed climate change as an existential threat to mankind and have implemented actions to eliminate fossil fuels from the generation of electricity. Some have claimed that wind and solar and other renewables can replace all the coal-fired, natural gas, and nuclear power plants in the United States. It can be argued that the actions these people are taking are making electricity more costly and less reliable, and placing Americans at risk for little or no reason. They are willing to gamble the safety and lives of Americans, as well as the American economy, on an ideology.

Our nation has suffered through a medical war fighting COVID-19 in which thousands died. As my neighbor said,

> The inability of our country to anticipate the coronavirus pandemic and put in place adequate reserves of all of the things we needed—PPE, ventilators, masks, tests, hospital beds, etc.—speaks loudly and directly to the need for reliable on-demand electricity and the need to plan for it right now.

Imagine if Americans had to suffer through rolling blackouts while quarantined at home during a future pandemic. How would newly erected emergency hospitals operate without electricity, let alone our existing hospitals without diesel fuel or natural gas to power emergency generators?

This was brought home by an oped in the *Washington Post*. Quoting from the op-ed:

> Residential use is up as workers and school children stay home.
>
> [Demand is down] in locked up restaurants, offices and factories.
>
> Hospitals are a different story: They consume twice as much per square foot as hotels . . . lead schools and office buildings by an even greater margin. And their work couldn't be more vital as they confront the novel coronavirus.
>
> A grid operator, sequestered in his dispatch center in East Greenbush, New York, said it all, "Keeping the lights on. . . . It's so critical."[3]

There is little doubt there will be another pandemic. The only question is when. We must do what is needed to guarantee adequate and reliable supplies of electricity in preparation for the next pandemic.

President Trump recognized the vital importance of the grid when he issued an executive order on May 1, 2020, to protect the grid from foreign adversaries. He said the grid, "provides the electricity that supports our national defense, vital emergency services, critical infrastructure, economy, and way of life."

There is also an ideology that threatens the grid. This book will examine how federal regulators, state governments, utility companies, and the operators of the grid themselves are imposing their beliefs about climate change on all Americans and placing the grid in great jeopardy. Unelected bureaucrats and self-imposed intelligentsia are making decisions that place all Americans in danger.

Looming Energy Crisis will show you why we must continue to use fossil fuels and why we must protect the grid from the actions of those who are imposing their personal beliefs on the rest of us. Our objective should be low-cost reliable electricity available for everyone.

Reliability is a national security issue.

Part One

Dangerous Goals

There's nothing more dangerous than
someone who wants to make the world a better place.

Banksy[4]

Chapter 1

The Grid

The average American has had no need to look behind the light switch to see how the massive electric grid works. What happens behind the wall on which the light switch is mounted has been the province of engineers.

If you are an engineer or have been involved with the operation of the grid you can skip this chapter. But, if you are like the vast majority of Americans who have never had to see how the sausage is made, you may find this very interesting. Alternatively, you may want to jump right to chapter 2, and refer back to this chapter when a question arises.

In many ways the grid is quite simple, but the details make it seem complicated. This review of the grid will keep it as simple as possible and only dig into details as needed.

The basic components of the grid consist of the equipment that generates the electricity, the equipment that carries the electricity across the country, and the distribution of electricity to the homeowner or business.

This book is about the grid—actually three grids: Eastern, Western, and Texas—that stretches across the United States, from the Atlantic to the Pacific Ocean. It's not about special situations that surround islands, such as Hawaii, or off-grid applications, such as might be found in isolated areas of the country. Islands lacking hydro and other resources, for example, have to import expensive fuels, such as diesel oil, which affects the cost relationship between fossil fuel powered generation and renewables.

We are blessed in the United States to have an abundance of low-cost fuels to power our grid, and to have the option of adopting renewables when they make economic sense.

Generation, Traditional

Traditional methods of generating electricity are listed first, along with appropriate details.

Coal-fired power plants

Most coal-fired power plants in the United States are of an older supercritical design having an efficiency of 33 percent. High efficiency low emission (HELE) plants are ultra-supercritical coal-fired power plants that operate at very high temperatures and pressures, made possible by advances in metallurgy. New HELE plants, such as those being built in China and Japan, have efficiencies of 45 percent. Only one HELE plant has been built in the United States because HELE plants emit slightly more than the 1,400 pounds of carbon dioxide (CO_2) per megawatt hour (MWh) permitted by the Environmental Protection Agency (EPA).

Natural gas combined cycle (NGCC) power plants

This combines a natural gas power plant with a steam turbine generator, using the exhaust gas from the natural gas unit to produce steam in a boiler to power the steam turbine. NGCC plants now have an efficiency of 63 percent.

Nuclear power plants

See appendix B for the current status of nuclear power plants.

Hydro power plants

These power plants are built in dams, and they produce very low-cost electricity. Most are built in conjunction with dams that provide flood control, but hydro turbine generators can also be added to smaller dams built along many rivers in the United States.

Traditional power plants, other than hydro, typically use two-pole generators that rotate at 3600 revolutions per minute (rpm), which produces three phase alternating current (AC), at 60 cycles per second (60 Hz), at 13,000 volts. Hydro uses multipole generators that also generate AC electricity at 60 Hz.

Generation, Renewable

Renewable power plants are listed next along with appropriate details.

Wind turbines

Chapter 7 provides details about wind power. One of the negatives of wind installations is that they kill birds and bats.

PV solar

Chapter 7 provides details about PV solar generation.

The output from wind and PV solar is direct current (DC) electricity that's converted to AC using inverters.

Concentrating solar

There are three types of concentrating solar.

The first uses mirrors to focus sunlight onto a boiler mounted at the top of a tower to generate steam to power a steam turbine. Ivanpah Solar Electric Generating System, in the Mojave Desert, California, is an example. Unfortunately birds are killed when they fly through the concentrated, focused beams of sunlight.

The second type of concentrating solar uses a horizontal reflector to focus sunlight onto a horizontal tube containing a fluid that can power a turbine generator.

The third type uses a parabolic reflector to focus sunlight onto a receiver that heats a fluid that drives a sterling engine to generate electricity.

Geothermal

Geothermal power plants are considered renewable, but can also provide baseload power. They use hot springs, i.e., geothermal energy, to produce steam, or a vaporized fluid, to power a turbine generator.

Geothermal power is limited because there are few locations in the United States where appropriate geothermal conditions exist. Iceland is the poster child for geothermal power because of the island's abundant thermal springs.

Transmission

Power transformers are used to raise the voltage to as high as 800,000 volts, and then reduce it to around 60,000 volts for sub transmission. High voltages reduce the losses incurred when transmitting electricity over long distances. Most AC transmission is done overhead, on towers, due to the very high cost of placing high voltage cables underground.

An alternative to AC transmission is to convert electricity from AC to DC. DC transmission reduces losses, but the equipment for converting AC to DC, and back again, is very expensive. DC transmission is useful for underwater transmission and for long distances, typically over 600 miles.

Distribution

Transformers in substations lowers the voltage to 13,000, 7,000, or 4,000 volts for distributing electricity within local geographic areas.

Distribution transformers, found on the telephone poles or in small boxes behind homes, lowers the voltage to 120/240, single phase, for use in homes.

Distribution lines can be put underground.

Other Equipment

Electricity is a powerful electro or electromagnetic force, especially at high voltages or currents. (Current measures the flow of electricity.)

Circuit breakers for interrupting the flow of electricity are extremely important. The small circuit breakers found in the home can interrupt the flow of electricity at low voltages. But at higher voltages, an arc is created when the flow of electricity is interrupted. This arc is hard to extinguish. At around 13,000 volts the arc is contained in a chute that allows the arc to be extinguished. At much higher voltages, huge circuit breakers standing several feet tall extinguish the arc using gases or other methods.

Distribution lines also use cutouts, a fuse, to interrupt the circuit if there is a short circuit on the line.

Voltages on distribution lines must be maintained within specific limits, and voltage regulators are mounted on the distribution lines to maintain the voltage within these limits.

Capacitors are mounted on distribution lines to correct for the effects, impedance, created by equipment using the electricity, which reduces the amount of useful electricity.

Reclosers are mounted along distribution lines to restore the flow of electricity when momentary short circuits occur. This typically happens when a tree branch falls across the lines creating a short circuit. Since the branch will most likely fall away, the short circuit is a momentary event. When this type of temporary event occurs, the recloser restores the flow of electricity and prevents shutting off the flow of electricity to the homes connected to this distribution line.

Important Considerations

Baseload power

Traditional methods of generating electricity, other than hydro, are built to operate 24/7, year round, and are referred to as baseload power. Attempting to have these plants follow load, with load increasing and decreasing, has negative effects of a varying nature.

Nuclear power plants can't be cycled up or down when demand varies.

Coal-fired power plants are damaged by the expansion and contraction of component parts when they try to follow load, due to the varying temperatures created with increasing and decreasing steam flows.

NGCC plants are more flexible and can adjust output more readily.

Natural gas peaking units, i.e., free standing gas turbine generators (without the boiler and steam turbine) are extremely flexible, and they can ramp up and down rapidly. Their purpose is to provide electricity during peak periods of demand.

Wind and PV solar, on the other hand, only operate when the wind is blowing or the sun is shining and can't be relied on to generate electricity when needed.

Storage

Storage has become an issue as attempts are made to use more wind and PV solar on the grid.

Pumped storage has been used since the early 1900s. It involves pumping water uphill to a reservoir when there is reduced demand for electricity, such as during the nighttime hours. Then, when there is a peak in demand, the water is released from the reservoir to flow through hydro turbines to generate electricity.

Recently, batteries have been used to store electricity, and this has become the method of choice since there are few locations available to construct dams that would allow for pumped storage. Lithium-ion batteries have become the battery of choice, though flow batteries have also been used. Lithium-ion batteries have to be replaced every ten years or so, which adds to the cost of storage.

Compressed air energy storage (CAES) is possible, with CAES installed at Huntorf, Germany, and the McIntosh Power Plant in McIntosh, Alabama, USA. The main drawbacks are too few locations for storage underground, or too expensive to create underground space by hollowing out geologic formations.

There have been numerous proposals for creating pumped storage, such as partitioning abandoned coal mines to allow water storage at different levels, or creating large holes in the ground at different levels for water storage.

Important Measurements

Capacity factor (CF) is an important measurement, for evaluating different types of generation. CF is determined by calculating the amount of electricity a power plant actually produces over a year, divided by the amount the unit should theoretically produce based on its nameplate rating.

Going by the nameplate rating alone can be misleading when comparing different methods of generating electricity. After all, it's the actual amount of electricity that's important.

CFs for the various methods of generating electricity are shown here:

- Nuclear: 91 percent

- Natural gas combined cycle: 87 percent

- Coal-fired power plants: 85 percent

- Onshore wind: For units built prior to 2016, 30 percent or less. Newer, taller and larger units, around 34 percent

- PV solar: 12–25 percent, usually based on location

Offshore wind hasn't been constructed in the United States, so CFs are not available. CFs in Europe, for offshore installations, generally vary from 35 percent to 48 percent.

Energy Transition

Proponents of climate change have demanded that fossil fuels no longer be used and that CO_2 emissions be eliminated.

These efforts to decarbonize the grid have raised serious issues.

The following chapters address these issues.

Chapter 2

Framework for Mandates

Several states and a significant number of electric utilities have committed decarbonization goals. The goals are frequently expressed as clean energy goals, in which clean energy is viewed as being synonymous with decarbonization.

Net-zero emissions is the term generally used when expressing decarbonization goals. The definition of net-zero emissions, as defined by the MyClimate website:

> Net-zero emission means that all man-made greenhouse gas emissions must be removed from the atmosphere through reduction measures, thus reducing the Earth's net climate balance, after removal via natural and artificial sink, to zero.[5]

This definition means that greenhouse gas (GHG) emissions are permitted as long as any emissions are removed from the atmosphere either naturally, e.g., forests, or artificially, e.g., carbon capture and sequestration (CCS). (Appendix A is a list of acronyms.) The concept of net-zero emissions is dishonest because the processes for removing CO_2 from the atmosphere are either not possible to measure or not actually doable. A fairer goal, and one that people can understand, is zero GHG emissions, where there is no fudge factor to hide behind.

Some utilities have established 100 percent decarbonization goals. This means they must rely entirely on wind, solar, hydro, and nuclear for the generation of electricity. But hydro is limited to a few states where there are existing dams with hydro turbines installed, while nuclear is being phased out and may not even exist by the end of this century (see appendix B).

There is also biomass (the use of organic material), but it's debatable whether it actually reduces CO_2 emissions. It assumes that the CO_2 released by burning biomass will be recaptured when trees or plants are grown to replace those that were burned. As a result, trees are being cut down, chopped up, and molded into briquettes. This is big business with substantial sales to Europe.

Other Questionable Biofuels

So they aren't overlooked, it's important to briefly examine other biofuels. There is algae, garbage, and woody pulp to make jet fuel. Garbage, for example, can be burned to generate electricity. An analysis of these alternatives is found in the book *Nothing to Fear*.[6] The primary obstacle in all these other biofuel and biomass proposals is that there isn't sufficient land area to grow or produce the feedstock to replace any significant amount of fossil fuels with biofuels.

Relying on wind and solar for 100 percent of electricity is a high-risk gamble. These sources also require storage, which adds substantially to cost. Unless one assumes that storage will allow wind and solar to provide 100 percent of our electricity, it's impossible to avoid using natural gas.

Every decarbonization goal and every clean energy mandate attempting to rely on wind and solar will, out of necessity, require using natural gas, with two possible exceptions:

1. There is the use of natural gas with carbon capture and sequestration.

2. There is the use of hydrogen in place of natural gas.

Natural gas, a fossil fuel, can be used to generate electricity if all the CO_2 that's emitted from the combustion process is captured and sequestered (CCS). Burning natural gas consumes the methane (CH_4) with the combustion process, resulting in CO_2.

It's been demonstrated that capturing CO_2 from a coal-fired power plant results in derating the plant by around one-third. In other words, a plant rated 300 megawatts (MW) becomes a plant rated 200 MW.

Derating is the result of using about one-third of the power plant's output to run the equipment required to capture and compress the CO_2 so it can be transported and sequestered underground. Derating of natural gas power plants could be harder to accomplish and require more power because the exhaust stream contains less CO_2, making it more difficult to capture.

As a result of the derating, it will be necessary to build at least four natural gas power plants instead of three—one extra that would otherwise not be needed to get approximately the same amount of electricity as before the derating. This will add to costs. (Example assumes 100 MW per plant.)

If natural gas is kept in the ground, for example by eliminating fracking, it would be virtually impossible to achieve any zero emissions goal. The exception would be to use hydrogen to replace natural gas, but this raises a myriad of issues.

Europe has proposed using hydrogen to replace natural gas and has put forth an extensive proposal to do so. The issue of using hydrogen in place of natural gas is thoroughly examined in appendix C.

In pursuit of clean energy, many states have established renewable portfolio standards (RPS) that utilities must follow (see figure 1). RPS require utilities to deliver to their customers a required quantity of electricity that comes from renewable sources as required by the state's RPS. These standards have mostly increased since they were initiated. A few are voluntary.

Subsequently, South Carolina has adopted an RPS goal (2014). Furthermore, certain states have mandated 100 percent of their electricity be clean by 2050. These states are in addition to utilities that have also committed to achieving 100 percent clean energy or zero emissions by 2050. Table 1, chapter 3, lists states that have clean energy mandates.

The RPS usually begins with a small requirement, say 5 percent, and then increases until it reaches a fairly large requirement by 2050. Figure 2 is an example of how a few states are increasing their RPS requirements between now and 2050.

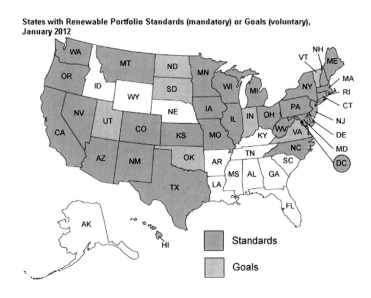

Figure 1. Map from EIA: States with portfolio standards.

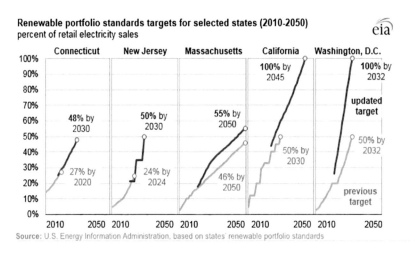

Figure 2. From EIA, shows some recent increases in RPS requirements.

The next two chapters describe additional clean electricity mandates and decarbonization goals.

Chapter 3

Clean Electricity Mandates

Clean electricity is a misnomer and misleading, yet it is the term being used by people who should know better, and it will be used here to be consistent. Table 1 is a list of twelve states with 100 percent clean energy mandates.

State	Date
California	2045
Colorado	2040
Connecticut	2050
Hawaii	2045
Maine	2050
Nevada	2050
New Jersey	2050
New Mexico	2045
New York	2040
Virginia	2050
Washington	2045
Wisconsin	2050

Table 1. States with 100% clean energy mandates.

The District of Columbia also has a goal to reach 100 percent clean electricity in twelve years, while they don't actually produce any electricity except for a few PV solar rooftop installations.

Mandated Outcomes

These twelve states listed in table 1, plus any others that adopt 100 percent clean electricity goals, have unwittingly mandated three outcomes:

1. Substantially higher costs for electricity for every resident.

2. Less reliable service with more frequent interruptions.

3. Increased risk for extended blackouts.

While a few of these states can benefit from hydro, most cannot, and will have to rely on wind and solar combined with storage.

<u>Substantially higher costs</u>

Using electricity that is only generated by wind and solar requires storage. Even if the cost of electricity from wind and solar is less costly than electricity generated by natural gas, which it isn't, the added cost of storage results in everyone paying much more for their electricity.

Remember, storage must store enough electricity for several days for when the wind doesn't blow or the sun doesn't shine. Added to this is the cost of building the excess wind and solar capacity to be able to generate the surplus needed for storage.

While it's not easy to estimate the cost for a state or city, the cost for the entire country to replace all fossil fuels is nearly $20 trillion, about equal to our national debt.[7] Plus, most of this cost is repeated every ten years as batteries and equipment wear out and have to be replaced.

The increased cost of electricity will be substantial. Germans already pay four times more than the average American, and they have only cut their CO_2 emissions by 31 percent. Imagine how much more they will pay when they start to add more than token amounts of storage.

More frequent interruptions

Batteries used for storage will fail and even catch fire. This will increase the number of interruptions because interruptions have rarely been caused by fossil fuel power plants. Nearly all the interruptions that occur today are because of transmission and distribution system failures, e.g., cables, transformers, flooding, fallen poles, etc. Batteries merely add additional potential points of failure.

Increased risk of blackouts

It's been proven that wind and solar are unreliable and can't produce electricity when certain conditions exist. Wind can't generate electricity when the wind doesn't blow or blows too hard, or if the temperatures get far enough below freezing that the turbines must be shutdown. Solar can't generate electricity when the sun doesn't shine or if snow and ice cover the PV solar panels.

If sufficient supplies of electricity haven't been stored in batteries, or elsewhere, there will be blackouts. No one knows how much storage is needed to prevent blackouts, or to support the community during a lengthy blackout before the wind starts to blow or the sun starts to shine. What happens if we guess wrong?

These risks are the result of 100 percent clean electricity, with no natural gas. If more states join the march to 100 percent clean electricity, costs and risks will increase.

And then there are the actions taken by individual utilities within many states. Some utilities have taken it upon themselves to establish decarbonization goals that either add to or complement any goals established by states.

Chapter 4

Utility Decarbonization Goals

Utilities have established decarbonization goals in response to state imposed RPS and state mandates, as well as on their own initiative.

Some companies are joining this game of decarbonization. Companies, such as Amazon, are also pledging to only buy electricity that is clean. What follows is a list of utilities that are all-in, with net zero or 100 percent clean energy by 2050.

- **Consumers Energy,** net zero 2040

- **DTE Energy,** net zero 2050

- **Consolidated Edison,** 100 percent clean by 2050

- Omaha Public Power District, net zero 2050

- **Public Service Electric & Gas,** net zero 2050

- Public Service Company of New Mexico, 100 percent emissions free 2050

- Austin Energy, net zero 2050

- **Dominion Energy,** net zero 2050

- Green Mountain Power, 100 percent carbon free by 2025

- Avista Utilities, 100 percent clean by 2045

- Madison Gas and Electric, net zero 2050

- Arizona Public Service, 100 percent carbon free by 2050

- Los Angeles Department of Water and Power, net zero 2050

- SMUD, net zero 2040

- Idaho Power Company, 100 percent clean 2045

- **Duke Power,** net zero 2050

- Platte River Power Authority, 100 percent non-carbon 2030

- **Xcel Energy,** 100 percent carbon free 2050

Additionally, there are forty utility companies with other goals ranging from 40 percent clean by 2035 to 80 percent by 2050. With fifty-seven utilities in the United States having goals to dramatically cut CO2 emissions, electric service will be affected across the United States.

Seven of the ten largest utilities by number of customers served (in bold type above), have decarbonization or clean energy goals of 100 percent by 2050. The remaining three of the ten largest utilities also have large decarbonization or clean energy goals.

- Next Era (Florida Power), third largest utility: Its goal is to achieve a 67 percent reduction in GHG emissions by 2025.

- Southern California Edison, second largest utility: Its goal is to reach an 80 percent GHG reduction by 2050.

- Southern Company (Georgia Power), fifth largest utility: Its goal is to achieve low to no GHG emissions by 2050.

These utilities blanket the United States, so vast numbers of Americans are directly threatened by these goals. The fact is, it's impossible for these goals to be met if the goals are taken at face value.

The "net" in net zero affords these utilities a loophole they can fall back upon when questioned about the veracity of their goal.

- "Net" allows them to claim they will offset their emissions by planting trees somewhere in the world, etc.

- "Net" also allows any utility, when questioned about its use of natural gas, to say it will eventually replace natural gas with green hydrogen or use CCS.

Both of these answers are bogus for a variety of reasons. The fact is, these utilities cannot meet their goals. Merely attempting to meet these goals will cause unnecessary hardships on Americans, resulting in:

- substantially higher costs for electricity for every American,

- less reliable service with more frequent interruptions; and

- increased risk for extended blackouts.

Chapter 5

What Will You Do in a Blackout?

Blackouts are all about life and death:

- People trapped in elevators

- Operating rooms without backup power supplied by diesel generators

- Trains stopped between stations, underground

- Airports without landing lights

- Food supplies interrupted due to lack of refrigeration

- Personal oxygen supplies exhausted

- Inoperable dialysis equipment

- Inoperable printing presses

- Inoperable cellphones

- Inoperable computers

- Inoperable TVs

While inoperable cellphones, computers, and TVs may seem trivial, how will people communicate? Since cellphones can't be charged when there is no electricity, how will people know if they are safe or what to do in an emergency? How will they contact their loved ones?

For a variety of reasons, natural gas and diesel fuel won't be available for emergency generators at hospitals or in homes. If the sun doesn't shine, PV rooftop solar won't work either. No one knows how long a blackout might last. Is one day's storage enough? Will the wind start blowing and the sun start shining in two days? Or one week? Or . . . ?

Blackouts are about reliability, and reliability is in doubt with wind and solar, which must have a means of storage.

Cost of Reliability

How much are you willing to pay for reliability? Li-ion batteries, the type being proposed for storage, currently cost $200 per kilowatt-hour (kWh). Their cost is projected to come down to $100/kWh by 2030. Meaningful storage is measured in MWh, or 1,000 times more. For reference, the Australian 100-MW battery supplied by Tesla was estimated to have cost around $100 million dollars. Its storage capacity was estimated to be around 130 MWh.

The real cost of electricity from wind and solar is far greater than the cost of electricity generated by NGCC power plants when the cost of storage is included . . . and storage is essential without the use of fossil fuels. (Appendix C examines electricity costs.)

Ten years ago, it was a fairly simple task to obtain reliable and factual data on the cost of generating electricity. Today, it is no longer possible to get this information from the government.

The Energy Information Administration (EIA) used to provide current levelized cost of electricity (LCOE) for each of the main methods of generating electricity. The EIA no longer provides actual, real-time information. Instead, it provides estimates for what the LCOEs will be in 2022, 2025, and 2040. These are estimates based on computer projections using data provided by bureaucrats who may have hidden agendas. Once again, bureaucrats have their thumbs on the scales. Even so, the best they can do is make it appear as though wind and solar have about the same costs, i.e., LCOEs, as NGCC power plants.

Table 2 is extracted from the February 2020, EIA report.[8] Note the LCOEs are for the year 2025.

Plant Type	LCOE $/MWh
Natural Gas Combined Cycle	36.61
Onshore Wind	34.10
Offshore Wind	115.04
PV Solar	32.80

Table 2. Plant type / LCOE.
Estimated weighted average LCOE for units entering service in 2025.

Zero emission and decarbonization goals and targets are costly and dangerous: blackouts kill. The federal government, including the Department of Energy (DOE), the Federal Energy Regulatory Commission (FERC), and the North American Electric Reliability Corporation (NERC) should be investigating electric utilities to determine how they intend to operate without fossil fuels.

Congress should insist on this.

We should insist on this.

Our lives are at stake.

Chapter 6

Out-of-Market Support

Suppliers come together in the marketplace to sell their goods. The auction is the marketplace for selling electricity. Events and actions outside this marketplace are having an important effect on which sources of electricity are actually being used on the grid. (The effect of subsidies, an out-of-market action, will be shown in chapter 10.)

Some states have also created carveouts and credit multipliers so as to increase the diversity of renewable energy supply. These are intended to prevent a single source, such as wind, from meeting all of a state's RPS requirement.

- A carveout specifies how much electricity must be supplied by a specific source, wave power for example.

- A credit multiplier results in multiple renewable energy certificates being awarded for a specific project. (See appendix E for greater detail.)

Renewable energy credits (RECs) have an economic value as they can be sold to other utilities who have not met their RPS requirements. Wikipedia states:

> Because RECs provide an additional revenue stream to renewable energy projects, they are essentially a subsidy meant to allow clean resources to economically compete with fossil fuel based resources.[9]

RECs are like printing money, and they are in addition to government financial subsidies. RPS is another out-of-market support.

Imposing RPS standards results in excess capacity, higher costs for consumers, and a degradation in reliability. It's obvious that costs will be higher as more costly wind and solar are added to the grid.

While costs can be debated (see appendix D), no one can deny that larger amounts of wind and solar will require storage, or that storage increases costs. Wind and solar are inherently unreliable and require costly storage to offset intermittency and unavailability.

California utilities are already adding storage because of the effect that relatively small amounts of wind and solar, less than 30 percent, are having on the grid. (Figure 7, chapter 11, establishes why storage is necessary to reduce the effect of sudden ramping of conventional supply in the evening.)

Out-of-market state support for renewables, including the imposition of RPS, increases costs and adversely affects reliability.

Chapter 7

Fundamentals of Wind and Solar

Since wind and solar are the primary means for replacing fossil fuels for generating electricity, this chapter will describe some of their fundamental quirks.

Wind

Wind turbines that have been installed until now are mostly rated 1.5 MW, 2.0 MW, and 3.0 MW. A few rated 5 MW have also been installed and a 10 MW unit is under development.

It's important to recognize that unlike natural gas, coal-fired, and nuclear power plants, wind turbines produce far less electricity than their nameplate ratings. This is because the wind doesn't blow consistently. Until 2019, most wind turbines had capacity factors of 30 percent or less. The newer, larger units, installed where the wind conditions are excellent, may have capacity factors of around 35 percent. Capacity factor (CF) is the amount of electricity a wind turbine, or any other power generation method, produces over a year, compared with how much it should produce using its nameplate rating.

The following gives energy sources and their typical capacity factors:

- PV solar: 12–25 percent

- Wind: 32 percent

- Coal: 85 percent

- Natural gas: 87 percent

- Nuclear: 91 percent

Figure 3 is a typical map of wind speeds at various heights aligned with the hub of a wind turbine. The capacity factors of wind turbines typically increase with the hub height where wind speeds are greater. They also increase with longer blades, which increase as the height of the wind turbine increases.

The output of the wind turbine increases slowly as wind speed increases. Typically, no electricity is produced until the wind speed is over 5 miles per hour (mph). The unit is usually most efficient when the wind speed is around 35 mph. When the wind speed goes above 55 mph, the turbine must be shut down to prevent it from being damaged. Similarly, the unit must be shut down when the temperatures go much below 0 degrees Fahrenheit (°F).

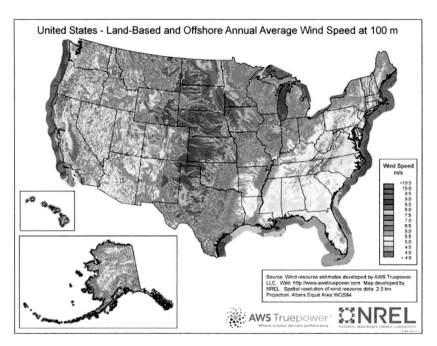

Figure 3. Map of areas where wind speeds are measured at 100 meter hub height.

The darkest areas in the middle of the continent and offshore, along the coasts, are where wind speeds are best.

An unfortunate fact about wind turbines is that they kill birds and bats.

Solar

There are two primary types of solar power plants: photovoltaic (PV) solar and concentrating solar

The most prevalent type of solar is PV solar. Panels with PV cells are mounted so as to best capture the sunlight. The panels can be mounted on rooftops and on supports in open land.

The amount of sunlight that reaches the Earth is measured as watts per meter squared. These insolation levels vary with latitude. The higher the insolation levels, the more electricity PV units can produce.

Figure 4 is a map of insolation levels around the United States. They are greatest in the Southwest and lowest in the Northeast and far Northwest. The amount of electricity that can be produced depends on how the PV solar panels face the sun. The most efficient are movable panels that constantly face the sun as it crosses the sky.

Concentrating solar is the other type of solar power plant. It comes in three versions. The most dramatic version is where thousands of mirrors focus the sunlight onto a boiler mounted at the top of a tower, which produces steam that drives a steam turbine generator. So far, these units are proving to be inefficient, even though they may include concentrations of molten salt where heat can be stored so as to be able to generate electricity after the sun sets.

Two other types of concentrating solar either use a parabolic reflector to focus the sunlight onto a collector at the parabolic focal point, or a tube-like reflector that focuses the sunlight onto a horizontal tubular collector.

PV solar is the most prevalent at this time. In fact, the 110 MW Crescent Dunes concentrating solar power (CSP) plant in Nevada has been shut down. Crescent Dunes was never cost effective, with electricity selling for 13 cents per kWh. When it had a catastrophic failure of its salt tanks, the purpose of which were to allow it to generate electricity after the sun set, the DOE issued a default notice and NV Energy, which had a purchase agreement with Crescent Dunes, terminated the agreement.

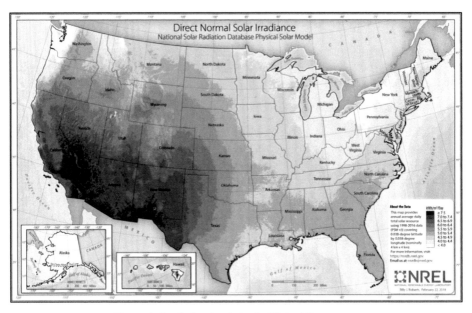

Figure 4. Insolation levels in the United States.

Gas turbine peaker units, which only operate during periods of peak demand, are being replaced with batteries charged by PV solar installations. While the gas turbine peaker can operate indefinitely, the batteries replacing the peaker unit can only operate for a finite period before running out of juice. This creates an unnecessary risk.

Part Two

Auction Manipulation

You can fool all the people some of the time
and some of the people all the time,
but you cannot fool all the people all the time.

President Abraham Lincoln[10]

Chapter 8

Market Structure

Power companies that operate in multiple states are involved in interstate commerce and fall under the jurisdiction of the FERC.[11] The NERC is responsible to FERC for ensuring the reliability of the grid.[12]

At the direction of FERC[13] in 1999, regional transmission organizations (RTOs) and independent system operators (ISOs) were established covering two-thirds of the population of the United States. The precise distinction between RTOs and ISOs is complicated, but they are fundamentally responsible for regulating the wholesale and retail markets within their assigned areas.

Figure 5 shows the locations covered by RTOs and ISOs.

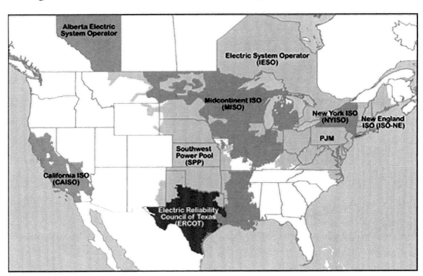

Figure 5. Map of regional transmission organizations (RTOs) and independent system operators (ISOs).

The remaining population, not covered by RTOs and ISOs, is covered by traditionally regulated utilities.

Prior to 1999, all of the utility companies were regulated by public utility commissions (PUC). Individual utilities, mostly vertically integrated, proposed prices, and also proposed what they considered as necessary investments in power generation and other facilities. The PUC reviewed these proposals and approved them, or as was generally the case, modified them.

Utilities made the key decisions as to how much and what type of capacity should be built, and what the prices for electricity should be for residential, industrial, and government consumers. They did this with PUC oversight, and continue to do so in one-third of the country.

While PUCs still exist in most states, and still approve rates and impose other requirements, the day-to-day operations controlling the grid, such as obtaining electric supplies from utilities and ensuring that adequate capacity is available for the future, resides with the RTO/ISOs in two-thirds of the country.

While our focus in part two is on FERC and the RTO/ISOs, there are other market conditions that need to be recognized in any thorough discussion of electricity markets. These are discussed in the next few paragraphs.

Competitive Retail Markets

The introduction of competitive retail markets for electricity, in which retail customers can choose their suppliers, further complicates the jurisdiction over market control because the states that adopt competitive markets are not necessarily within RTO/ISO areas. Retail customers in competitive markets can select, for example, a supplier who distributes electricity from renewable sources.

Virtually all of the competitive retail markets are in states covered by RTO/ISOs. Georgia is an exception.

Power Purchase Agreements

Power purchase agreements (PPA) are where independent owners of electric generation facilities, i.e., power plants, sell electricity under contract to consumers, utilities, and RTO/ISOs. These agreements are for twenty years or so, which guarantees the builder of the power plant—typically wind or solar facilities—that costs will be covered and financing can be secured.

Ancillary Services

There are other services that must address such items as frequency control and reactive power. These ancillary services are also covered by various payment mechanisms, but the aggregate of these payments are relatively small. For our purposes, we will slice through this complexity by focusing on how the markets for electricity are controlled in areas under the jurisdiction of RTO/ISOs. This is different from those states that operate traditionally regulated markets where regulators approve the price of electricity and the availability of so-called green electricity.

The formation of RTO/ISOs is likely the result of deregulation efforts such as those started by President Carter with the deregulation of the airline industry (1978 Airline Deregulation Act). Deregulation was an effort to introduce market forces into what were considered to be inherently monopolistic industries.

Electric utilities were inherently monopolistic. Utility companies had been granted the right to serve specific areas under the scrutiny of regulators. The regulators were responsible for ensuring utilities would get a fair rate of return on their investments without overcharging customers. In addition, the regulators determined whether investments proposed by the utility were necessary. This defines traditional regulated markets.

Some people believed that traditionally regulated markets restricted competition and prevented new sources of electricity from being introduced into the market. The breaking up of AT&T was the historic precedent. (In 1974, the U.S. Department of Justice opened up a case against AT&T. The case was finalized in 1982 by the mandated breakup of the Bell System.)

The impetus behind RTO/ISOs was the belief that competitive market forces could be introduced by replacing "regulated" markets with "competitive" markets operated by RTO/ISOs. Ostensibly, RTO/ISOs would conduct their operations in a way that would result in the lowest cost of electricity for consumers while maintaining reliability.

It should be noted that RTO/ISOs cannot own power plants, but they merely establish the rules for how power plant owners will operate. It's also important to note that not all RTO/ISOs are the same.

California Assembly Bill 1890, the Electric Utility Industry Restructuring Act, made it clear that:

> [t]he proposed restructuring of the electricity industry would transfer responsibility for ensuring short- and long-term reliability away from electric utilities and regulatory bodies to the Independent System Operator and various market-based mechanisms.[14]

Note that this shifted responsibility for reliability from utilities to ISOs. However, we will establish that the integrity and reliability of the grid as well as the low-cost objective have been jeopardized by politicians at both the state and federal levels who have been pursuing policies based on political agendas.

The fear of climate change, with the objective of cutting GHG emissions, is central to these political agendas. In essence, RTO/ISOs have rigged the market to ensure that wind and solar replace fossil fuels.

Prior to the emergence of deregulation and subsequently RTO/ISOs, it was generally believed that the public was best served by a utility industry that was essentially monopolistic. There was agreement that installing duplicate transmission and distribution lines, or establishing duplicate excess power generation capacity, would result in higher costs and inefficiencies that would not be in the public interest. Until now, there has also been agreement that the utility industry must provide electricity at low-cost and that the supply of electricity must be reliable.

Millions of words have been written about these complex issues by academics, lawyers, and technicians using terms most people cannot understand. While it's true that the issues are extremely complex, it is still possible to focus on how the system works and how politics are distorting the system in ways that harm most Americans. This is an attempt to boil down the complexities to show how and why the system is being manipulated to America's detriment.

The system has become politicized, with those who believe that decarbonation must be accomplished opposed by those who believe that climate change is not an existential threat and favor policies that achieve low cost and greater reliability.

FERC consists of up to five commissioners, with no more than three from the same party. With Democrats generally favoring decarbonization, and Republicans generally favoring low costs and greater reliability, FERC has become divided.[15]

A regulated system covers one-third of the population, while RTO/ISO organizations cover two-thirds of the population. As a result, while the country operates under two systems, FERC is ineffective. While FERC and state regulators are fairly transparent, RTO/ISOs are very opaque.

Unraveling how the RTO/ISOs operate is a huge challenge. Are they striving for low costs? Or are they imposing their beliefs about climate change on the people they serve?

Chapter 9

RTO/ISO Process

RTO/ISOs conduct auctions in an apparent attempt to instill competitiveness into the process of deciding (1) which producer's electricity is fed onto the grid and (2) which new power generation power plant is built to meet future needs. The day-ahead and real-time auctions are used for deciding which producers' electricity is fed onto the grid. Capacity auctions are used for determining which new power plants are built to meet future demand. This chapter presents an overview of how RTO/ISO auctions operate.

Each RTO/ISO must ensure the availability of electricity at all times and uses the following process to meet that objective:

- Each RTO/ISO attempts to enter into agreements that ensure the availability of electricity daily, within their areas, while also ensuring there will be sufficient future generating capacity.

- Before using auctions, the RTO/ISOs enter into PPAs with independent operators for some of their daily needs. Agreements to purchase hydro power, when it's available, can also account for a portion of each day's needs. Some states, within RTO/ISOs, have mandated the use of nuclear power, and this also provides a portion of the day's needs. These agreements, PPAs, etc., provide a portion of the electricity needed each day.

- RTO/ISOs then conduct a day-ahead auction involving all power plants within their jurisdiction and attempt to schedule delivery of specific quantities of electricity for specific periods, say during five-minute intervals, for the next day.

- On the next day, they conduct real-time auctions to fill in any gaps that occur during the day that the day-ahead auction has failed to make available.

The combination of contract purchases, day-ahead, and then real-time auctions are designed to meet the demand for electricity during each and every day.

The system must comply with a few unalterable truths. For example, there must always be sufficient supply to meet demand, or the system collapses. Frequency, in cycles per second, i.e., Hz, must never be allowed to deviate by more than a fraction. User voltages must be maintained within a small range.

Some of the RTO/ISOs then conduct capacity auctions, the results of which are supposed to ensure there will always be sufficient capacity to meet future demand.

The first flaw in this complex system begins with the day-ahead auction.

Chapter 10

Rigged Auctions

This chapter will establish why the RTO/ISO systems are rigged in favor of wind and solar. Subsequent chapters will describe why the wrong bidding system is being used for day-ahead auctions and why the current bidding system is inappropriate. Other chapters will delve into why the capacity auctions are becoming inappropriate and how they are being manipulated to include wind and solar.

It's becoming clear that the RTO/ISOs are imposing their beliefs about climate change on the people they serve by using inappropriate auction systems.

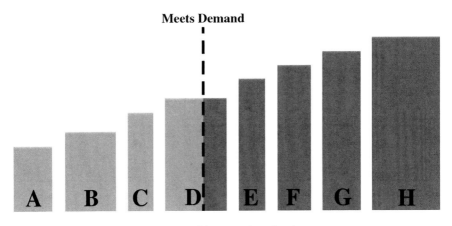

Figure 6. Bidders in a day-ahead auction.

Figure 6 is a chart showing various bidders from A to H who participate in a day-ahead auction. We will use figure 6 to explain why the day-ahead auction is rigged in favor of wind and solar.

Bidding progresses until bidders supply enough electricity to meet the forecast demand. In this case, bidders A, B, C, and D submit bids that meet all the forecast demand, and in so doing, clear the market.

Let's distinguish the bidders, reading from left to right, i.e., A to H:

- A, B, and C are bidders that cleared the market and are supplying electricity to the grid.

- D is a bidder that partially cleared the market and is providing some electricity to the grid.

- E through H are bidders that did not clear the market and are not providing electricity to the grid.

Let's make the following assumptions:

- Bidders A, B, and C are from wind and PV solar suppliers.
- Bidder D is from an NGCC supplier.
- Bidder E is from a nuclear power plant.

Bidders A, B, and C

Wind and PV solar suppliers receive other payments or grants to cover their costs and can, as a result, bid as low as $0, though they may bid somewhat more.

Because of this, wind and PV solar are always the winning bidders as long as wind or solar is available (for example, if the sun isn't forecast to shine, there may not be a PV solar bidder).

Bidder D

Bidder D, in our example, is the marginal resource or price setter and gets paid 6 cents per kWh, the amount it bid, for the electricity it produces.

Typically, when there are no more wind or PV solar bidders, the next bid will be from an NGCC supplier (such as bidder D) because its bid is usually lower than bids from nuclear power plants.

Bidder E

Bidder E, a nuclear power plant, can't currently compete against NGCC plants because of the low cost of natural gas, or against wind or solar plants that receive subsidies.

Bidders F, G, and H

These bidders could be from coal-fired power plants or from other NGCC or nuclear power plants.

Auction

Let's look at a hypothetical auction:

- Bidder A bids 0 cents per kWh.

- Bidder D bids 6 cents per kWh. (Bidder D is the price setter.)

- Bidders A, B, and C also get paid 6 cents per kWh.

Why are bidders A, B, and C being paid more than they bid?

Wind and PV solar providers merely need to ensure they bid below any other bid to guarantee they still get paid whatever the marginal resource or price setter bids. And they can nearly always bid lower because they have other sources of income (other payments, tax credits, or grants).

Wind, for example, gets paid the production tax credit of 2.3 cents per kWh for every kWh they produce. They just need to be certain to win the bid so they can generate the electricity for which they receive the 2.3 cent per kWh subsidy.

The amount of the production tax credit varies depending on the year the wind installation began construction and extends for ten years, beginning with when the facility starts generating electricity.[16]

Consequences of Auctions

Real-time auctions operate in a similar manner to day-ahead auctions. Both have three immediate consequences:

First, they result in nuclear and coal-fired power plants being retired, which reduces the availability of baseload power. Once a coal-fired or nuclear power plant is retired, it can't easily be replaced. Essentially, it is gone forever.[17]

Second, they encourage the building of unnecessary power plants. Wind and solar plants are built because they get subsidies and can always win these auctions, even though there may not be a need for the electricity they produce other than to meet renewable portfolio standards required by some states. Utilities and independent power producers are building these wind installations to take advantage of the tax credits.

Third, the levelized cost of electricity from wind and solar is higher than from NGCC, coal-fired, and nuclear power plants. While this statement is controversial, costs are discussed in greater detail in appendix D.

Why should all the winning bidders be paid the same price? Wind and solar have a competitive advantage because they can bid zero and still have their costs covered by an out-of-market payment—a payment hidden from the bidding process.

This is skewing the bidding so that wind and solar can always win any bid they enter.

Chapter 11

Uniform Clearing Price Auction

The bidding process described in chapter 10, is known as a uniform clearing price auction (also called a single clearing price auction) and is common in auctions involving commodities. However, this type of auction isn't appropriate for the electricity market.

A true commodity, such as wheat, is identical whether it's grown in Iowa or Nebraska. Bushels of wheat are interchangeable. At first blush, it would seem that the same would be true with a barrel of oil. But the composition of the oil in the barrel varies between locations, such as whether it's extracted from the North Sea or from shale in Texas.

For this reason, the price for oil will vary depending on its source. Brent establishes the price for oil from the North Sea; West Texas Intermediate is the price for oil from the Permian Basin in Texas; Dubai Crude is extracted from Dubai; and so forth. While oil appears to be a single commodity, it actually varies by source.

Upon first impression, electricity would seem to be a single commodity since the electrons that make up the flow of electricity are all the same. Some providers of electrons can guarantee their delivery twenty-four hours everyday of the year (while others cannot). These electrons are more valuable than electrons that may or may not be available when needed. Economists assign values based on time and place, and guaranteed availability has value.

The problem with attempting to use the uniform clearing price auction in the electricity market is that electricity from wind and solar isn't available 24/7. There's no guarantee it will be available when needed. While seemingly subtle, the difference is huge.

A natural gas combined cycle, nuclear, or coal-fired power plant can guarantee electricity is available when needed, which is one reason why they are known as baseload power plants.

It's actually impossible for electricity from wind and solar to be available 24/7 unless sufficient storage is available to store enough electricity generated by wind or solar to make it available 24/7—and that cost is huge and not included in the auction. The suppliers of electricity from wind and solar should have to guarantee they can supply the electricity when needed, and to do that they would have to include the cost of storage.

But, the providers of wind and solar don't usually provide the storage. Storage is usually supplied by the utility. Until now, there's been little need for storage since wind and solar have supplied only small amounts of electricity. With wind and solar accounting for a greater portion of supply, the need for storage has become essential.

The California Independent System Operator (CAISO) identified this need a few years ago with a graphic called the Duck Curve. CAISO recognized the need for storage to minimize the sudden ramp up when the sun sets. In addition, CAISO identified the need for storage to accommodate an oversupply of electricity during the day, when solar generates more electricity than there is demand. It also recognized that the problem would get worse as wind and solar supplies increased.

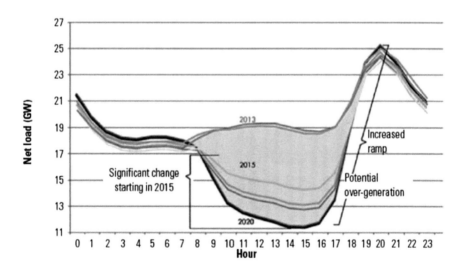

Figure 7. CAISO Duck Curve.

The uniform clearing price auction used by the RTO/ISOs is not suitable for determining the lowest cost bidders because an increasingly large part of the cost of wind and solar are excluded and because the value of a guaranteed supply is ignored. This problem can be partially solved when conducting auctions by only paying wind and solar bidders what they bid, rather than paying them what the winning bid gets paid in uniform clearing price auctions, and then also requiring the wind and solar bidders to pay for the necessary storage or backup.

This would tend to force politicians to increase the subsidies paid to wind and solar suppliers because wind and solar would otherwise be losing money, which would put the problems caused by wind and solar where it belongs, at the feet of politicians.

Chapter 12

Supply Must Meet Demand

One aspect of the grid that many people may not be aware of is that the supply of electricity must equal demand. If supply doesn't meet demand, the grid will collapse. It's not possible for supply to *almost* meet demand: It must equal it at all times.

This is why the lack of a small amount of solar or wind can cause the grid to crash. If all of the supply from natural gas, coal, nuclear, or hydro is already available—and maxed out—and only 1 percent, or even less, of the supply from wind or solar becomes unavailable, the grid will crash.

A recent example is what happened in South Australia, where the last coal-fired power plant was closed. A storm caused the supply of electricity from wind to suddenly stop. As a result, the entire state of South Australia suffered a blackout for more than twenty-four hours.

Blackouts over wide areas for extended periods shouldn't be confused with short interruptions caused by the failure of equipment. An underground cable may fail in a community, which shuts off electricity to a group of homes or an area within a city. This is what happened July 13, 2019, on Manhattan in New York City.

Preventing Collapse

RTO/ISOs and utilities in regulated markets have only four possible methods to avoid such a crash if electricity from wind or solar suddenly becomes unavailable:

1. Basically, they can shed load or establish rolling blackouts.

   • The first step would be to lower the voltage on the system, which results in automatic load shedding.[18]

- Load shedding can resolve short term or minor disturbances where demand exceeds supply.

2. In the second instance, when demand substantially exceeds supply for longer periods, RTO/ISOs and regulated utilities will have established arrangements with customers who can shed load without hurting their operations.

   - The current term for this type of activity is "demand response," or DR, though the practice has been in place for decades. By acting quickly, the RTO/ISO or utility may be able to avoid collapse of the grid by shedding enough load so that supply again meets demand.

   - A chain store or manufacturer may enter into an agreement whereby they get a favorable rate for the electricity they use for allowing the RTO/ISO or utility to curtail the supply of electricity.

      ◊ A chain of supermarkets could shut off half the lighting in their stores to shed load. (Simplistically, a remotely operated breaker could cut off the electricity to the unessential lighting.)

      ◊ A manufacturer may be able to shut off non-critical lighting or air-conditioning to shed load.

3. The next, more drastic alternative, is for the RTO/ISO or utility to cut off supply to specific areas in their territory. These are referred to as rolling blackouts.

   - Rolling blackouts have been common in undeveloped countries, but not in the United States. The United

States will become just another underdeveloped country if it reaches the point where rolling blackouts become common.

- For example, the NYISO could cut off downtown New York City and cause a blackout in that area, and then restore electricity to that area later, after cutting an equal amount of demand elsewhere, such as cutting the supply of electricity to Brooklyn.

4. The fourth way to protect against the collapse of the grid when wind or solar fail to produce electricity is to have gas turbines standing by in spinning reserve, ready to come online at a moment's notice.

- This may not be possible during the winter in northern states where the supply of natural gas is unavailable to NGCC power plants because natural gas must go to homeowners for heating purposes before it can be used to generate electricity.

- There are some areas, such as in New York and New England, where state governments have prohibited the construction of pipelines, preventing the availability of sufficient supplies of natural gas in those areas.

- This problem is migrating to other areas of the country as state governments prevent the construction of pipelines to meet the growing demand for natural gas. Protesters are preventing the construction of pipelines by using the courts to block construction for a variety of reasons. Many of these pipelines are essential if there is to be a guaranteed supply of electricity in certain areas of the country.

- Storage would be the final alternative for providing electricity when the wind or solar supplies suddenly fail.

Just because wind and solar supply only a small percentage of the grid's electricity is no reason to assume that the sudden loss of supply from wind and solar can't cause the grid to collapse.

Chapter 13

Other Market Considerations

Each RTO/ISO must control other aspects of the grid with auxiliary auctions or procedures. Electricity flowing throughout the grid must be maintained at 60 cycles per second, i.e., 60 Hz. This frequency must be maintained within small limits or else the grid collapses.

Traditional generation (i.e., coal-fired, natural gas, and nuclear) inherently provide frequency stabilization within limits, while wind and solar do not provide frequency regulation. Inverters can help in this regard, but wind and solar complicate this aspect of grid management. RTO/ISOs are also responsible for establishing markets for frequency control.

Another factor that is very important is transmission. Transmission lines have limited capacity and must not be overloaded. If they are overloaded, they overheat. Overheating can cause lines to stretch and sag, potentially causing a failure and crashing the grid. In addition, transmission line constraints can affect distribution of electricity within each RTO/ISO.

Another factor that causes problems is that electricity doesn't flow from A to B in a straight line. Rather, it flows through multiple paths in difficult to predict ways; this can cause problems for managing the grid.

Each RTO/ISO is responsible for establishing auctions and otherwise controlling transmission lines to ensure the smooth flow of electricity within their regions. These are referred to as ancillary services.

Some observers have assumed that bids represent the marginal, or variable, cost of the supplier and make the argument that the auction provides the least costly electricity.

The price setter, which is the marginal resource in figure 6, is the bid that establishes the highest accepted price for each auction, but it doesn't mean that the price setter, or the other winning bids, reflect their marginal, or variable, costs.

In any business that produces a product, whether it be a car, a ton of iron ore, or a kWh of electricity, the next unit produced represents the marginal, or variable, cost, but only if all the fixed costs have already been covered by previous production or some other source of revenue, such as a subsidy.

Once all the fixed costs have been covered, the next unit of production only includes variable costs. The price bid in the auction may represent the full cost of producing the electricity, but could represent the variable cost if all the fixed costs have already been covered.

No business can survive if it doesn't cover all of its fixed costs; therefore, the price set by the winning bidders in these auctions is not necessarily the variable cost of producing the electricity being accepted for the grid.

Chapter 14

Capacity Auction

One of the responsibilities of RTO/ISOs is to ensure that there will always be sufficient supply. Capacity auctions are being used for this purpose by some of the RTO/ISOs. Bidders commit to building or retaining sufficient capacity to be able to guarantee adequate supplies of electricity over the next few years.

The specifics of capacity auctions vary between RTO/ISOs, but when used, are of the same type (i.e., uniform clearing price auction used for day-ahead and real-time auctions where the accepted bids all receive the same clearing price).

The basic structure of capacity auctions are organized in the following manner:[19]

1. The auction covers a specific future period, such as one or three years ahead.

2. The RTO/ISO determines the required capacity needed to ensure adequate supply for this time period.

3. The RTO/ISO establishes the price cap, which is the "cost of new entry" (CONE), or cost for building a new power plant, such as an NGCC power plant.

4. There is then a determination of the amount of money the new capacity will receive over its life from selling electricity, and this is the net-CONE. The net-CONE is subtracted from the CONE, which identifies the missing money for creating the new capacity.

5. Winning bids are determined from a bidding structure consisting of the net-CONE and CONE and a curve representing future demand.

Essentially, this approach accounts for the revenues a new plant will earn once it's built, so that the plant owner doesn't get paid twice.

Alternative Proposals

The purpose of these auctions is to ensure the availability of future capacity. But how can wind and solar installations guarantee they can provide electricity at any time in the future when it's impossible to guarantee that the wind will blow and the sun will shine?

Should wind and solar be excluded from capacity auctions? If wind and solar can't participate in capacity auctions, new wind and solar plants would have to be built speculatively. PPAs for twenty years provide a method for doing this.

The RTO/ISOs who have adopted capacity auctions have been struggling to arrive at a process that includes wind and solar where future availability is guaranteed. RTO/ISOs have proposed some alternatives, but FERC has the ultimate authority for approving how auctions are conducted, and the RTO/ISO alternatives have not always been approved by FERC.

These proposals supposedly establish that wind and solar power generation can be used to provide guaranteed future capacity. It's tantamount to claiming that RTO/ISOs can determine when the wind will blow and the sun will shine.

Capacity value

The PJM Interconnection (see figure 5; see appendix A) has developed a process for arriving at capacity value for wind. Through statistical analysis, the PJM has estimated that wind has an effective load carrying capability (ELCC) of 10–11 percent.[20]

PJM says, "ELCC is a measure of the additional load that the system can supply with the particular generator(s) of interest, without change in reliability."[21] The ELCC uses historic data to develop a resource adequacy case that meets the "one day in ten years" loss of load expectation (LOLE) criteria, and then uses historic data for wind and solar to be certain the one-day-in-ten-year requirement is met.

Historically, baseload power plants, i.e., NGCC, coal-fired, and nuclear, have supplied the grid, and these have formed the basis for the ELCC. But as NGCC, coal-fired, and nuclear power plants become no longer available, and wind and solar become the predominant or sole sources of electricity for the grid, the LOLE is no longer valid. As described in chapters 2–4, the decarbonization goal is to eliminate all natural gas and coal-fired power plants from the grid, while nuclear power will also disappear if new ones aren't built.

Assume, for example, there is no baseload power available and wind and solar are the sole suppliers of electricity, in which case the statistical underpinnings of the ELCC evaporates. Part of the formulation of the ELCC is that baseload power is available if wind and solar are not available. There is no 100 percent guarantee that capacity will be available from wind and solar when needed. Grid reliability is reduced by using this type of statistical approach.

MOPR and CASPR

The minimum offer price rule (MOPR) was introduced by at least one RTO/ISO to ensure that subsidized providers, such as wind and solar, couldn't lowball the capacity auction and prevent baseload power plants from winning the auction.

MOPR could exclude wind and solar from building future capacity. However, in ISO-NE, there was an exemption for a limited amount of wind and solar to bid in the MOPR auction. This was known as the Renewable Techology Resource (RTR) exemption.

To allow MOPR to be put in place, the ISO-NE proposed having a two-step auction where MOPR would be the first step and the Competitive Auction with Sponsored Policy Resources (CASPR) would be the second. This would lead to phasing out the RTR. Wind and solar can participate in the CASPR auction.

In approving CASPR, FERC said:

> The collection of revisions, referred to as Competitive
> Auctions with Sponsored Policy Resources (CASPR),
> adds a secondary auction to the Forward Capacity
> Auction (FCA) process to facilitate the transfer of ca-
> pacity supply obligations from existing capacity re-
> sources, which commit to permanently exit ISO-NE's
> wholesale markets, to new state-supported resources
> (Sponsored Policy Resources) . . .
>
> This out-of-market state support raises a potential
> conflict with the Commission's interest in maintain-
> ing efficient and competitive wholesale electric mar-
> kets.[22]

Note two aspects of this statement:

1. MOPR and CASPR allow winning bidders to sell their
   capacity commitments to renewable sources providing the
   winning bidder permanently exits the market.

2. State policies, such as renewable portfolio standards (RPS)
   regarding out-of-market, i.e., subsidized wind and solar,
   could result in countermanding the purpose of capacity
   auctions.

MOPR and CASPR would seem to undermine the purpose of
providing guaranteed capacity, especially when a winning bidder,
presumably from an NGCC power plant, can sell the contract to a
wind or solar provider. This would eliminate more baseload power
from the grid. This was FERC's response to the proposal from the
ISO-NE, but FERC has, as of this writing, not approved changes to
the PJM auction using capacity value.

While it's not possible to say how often the grid will fail for lack
of capacity when wind and solar are relied on to supply the capacity,
the fact they will not be available at some time means there is a great-

er probability of blackouts and that the grid is less reliable.

The greater the complexity, the greater the probability of failure. We are fast approaching the time when we will be facing blackouts.

Chapter 15

Reserve Margins

Another aspect of guaranteeing the future availability of electricity is reserve margins. Simply stated, the amount of capacity intentionally maintained above any prior peak demand is considered the reserve margin.

Reserve margins are supposed to ensure there will be sufficient generating capacity even if demand suddenly and unexpectedly surges higher than ever before. This could happen, for example, during a heat wave when the use of air-conditioning increases.

An RTO/ISO covering Texas, ERCOT, has seen its reserve margin become dangerously low, at 7.4 percent. In the past, its targeted reserve margin was 13.75 percent.

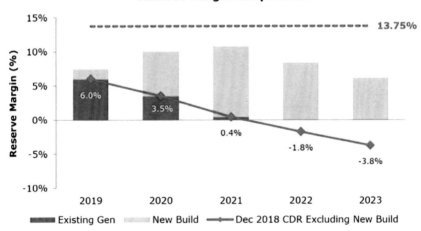

Figure 8. Texas reserve margins.
From NRG Energy, Fourth Quarter and Full Year 2018 Earnings Presentation.
CDR = Reserve Margin.

But, what is the reserve margin when wind and solar are added to the grid? How can they be included as a reserve for use in an emergency when the reserve might not be there? Power from wind might not be available if there are freezing temperatures, if it's too hot and the wind doesn't blow, or if the wind blows too hard. Likewise, power might not be available from solar when skies are cloudy for several days.

Figure 8 shows how much new build will be required in ERCOT's territory to ensure a safe grid without blackouts. But how much of the new build can be wind or solar while still maintaining a safe grid?

The sloping line in figure 8 is the reserve margin from existing baseload resources. For example, 3.5 percent is the reserve margin attributable to baseload power plants in 2020. But it turns negative in 2022 and 2023, when existing baseload power plants can no longer meet anticipated peak demand.

How safe will Texans be from blackouts when there are negative reserves, i.e., shortfalls from baseload power plants? If all of the new build is wind and solar, the existing baseload power plants won't be able to meet peak demand. The system would fail if wind and solar suddenly became unavailable. (Failure is when available power can't meet peak load.)

ERCOT must wrestle with the same issue as the PJM:

- How much of the new build can be wind and solar?

- When is reliability placed in danger?

- What about the following years?

When viewed from this perspective, the PJM statistical approach looks even less viable.

Whenever we deal with probabilities we must be willing to accept failure. Is failure acceptable? Are preventable blackouts acceptable?

The purpose of the reserve margin is to guarantee that electricity will always be available, barring equipment failures, so there won't be blackouts. Capacity auctions should take reserve margins into consideration when they establish the new capacity that's required.

Chapter 16

Real World Example

The Harvard Business School introduced case studies to allow students to obtain a feel for the real dynamics of how businesses operate. Similarly, a study of actual past events can help make clear why RTO/ISOs are failing to achieve their intended purpose.

New England Winter, 2018

During the coldest days of January 2018, the New England grid was on the brink of disaster. Only pre-planning by ISO-NE, emergency actions by the Coast Guard, and luck kept the lights on.

Here is the situation New England found itself in during the winter of 2018:

ISO-NE had required that oil supplies be established on site at NGCC power plants so they could burn oil if their supplies of natural gas were cut off. (Gas turbines can burn oil and other gases, such as hydrogen, if properly equipped to do so.)

The law requires that homeowners have first call on natural gas supplies, and power plants cannot be certain of having adequate supplies when cold weather results in increased loads.

Natural gas cannot be stored at the power plant and must be drawn directly from the pipeline.

No new pipelines can be built through New York State, so there is a shortage of natural gas for the New England states.

Oil was the fuel that kept the lights on in New England during January 2018.

As the cold weather worsened in December and January, oil supplies were rapidly depleted. Power plants were scraping the bottom of the barrel of their oil supplies. By January 9, natural gas power plants were down to a one-day supply of oil. Resupply was necessary.

Waterways were blocked by ice; therefore, ships carrying oil couldn't reach the power plants. The Coast Guard was called to open the ice blocked waterways with ice breakers so refueling vessels could reach the power plants in need of replacement oil. Bridges were raised during rush hour to allow these vessels to pass so the oil could reach its destination in time.

Luck asserted itself as the weather finally broke, just in time to allow normal operations to resume.

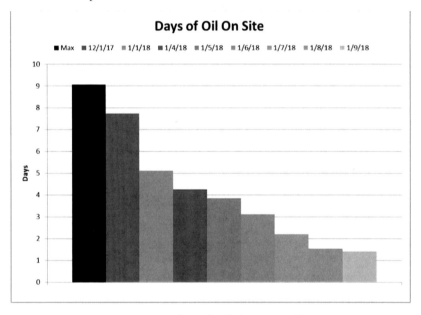

Figure 9. Oil supplies from ISO-NE.[23]

If the cold weather had lasted for one more day, there would have been blackouts. Wind and solar were not able to provide the needed amount of electricity to the grid. New England and the northeastern states are a poor location for PV solar any time of the year, but especially in the winter. The insolation level for Boston in January is 1.66 kWh/square-meter/day. (All future references to solar insolation levels use kWh/square-meter/day. Tables of insolation levels are readily available from the Internet.)

Figure 10 shows the average annual solar resource using tilt latitude collectors.

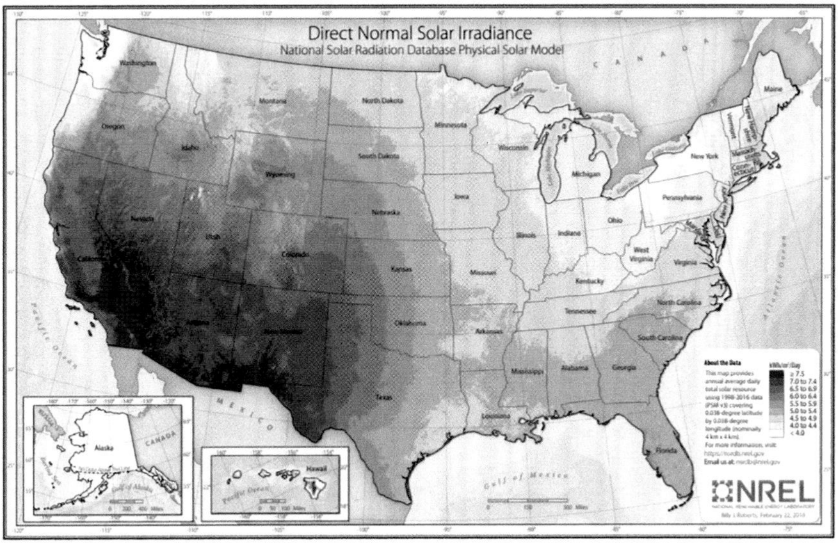

Figure 10. US solar resources, from NREL.

For comparison, the insolation level in Phoenix, Arizona, during January, is 3.25. During January in New England, an average insolation level of 1.66 is the best that can be expected, but it doesn't reflect the denigrating effect of snow covering PV panels.

The additional degradation caused by snow and other winter precipitation can be seen from these charts from the ISO-NE report for front-of-the-meter and behind-the-meter electricity produced by PV solar installations during January 2018.

The light-grey line in figure 11, on the next page, represents the anticipated power produced from PV installations in New England during the period from December 24 through January 9, 2018. The dark-grey line at the bottom shows the actual power generated during this period.

Snow cover is most likely an important contributor to this tremendous degradation. In New England on some days (e.g., December 25 and January 4), there was virtually zero electricity generated by the sun. For most other days, the amount generated could best be described as anemic.

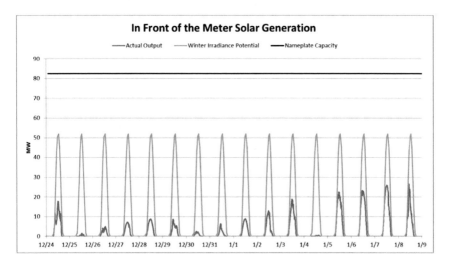

Figure 11. In front of the meter solar generation.
From ISO-New England Cold Weather Operations Presentation, January 2018.[24]

Figure 12, a chart for behind-the-meter installations, demonstrates essentially the same anemic results.

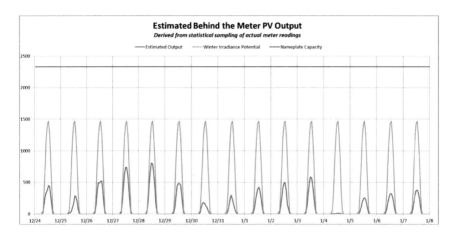

Figure 12. Estimated behind the meter PV output.
From ISO-New England Cold Weather Operations Presentation, January 2018.[25]

Wind turbine performance during this period is shown in figure 13. The top line is the predicted supply, the next lower line is the actual supply, while the bottommost line is curtailment.

Curtailments, in grey at the bottom of the chart, were primarily due to transmission congestion. Lack of output was mostly caused by intermittent shutdowns due to high wind speeds.

ISO-NE may be one of the most endangered RTO/ISOs in the country—with the possible exception of CAISO, where state policies are driving the grid to an untenable condition.

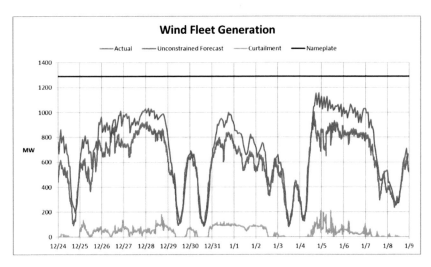

Figure 13. Wind fleet generation.
From ISO-New England Cold Weather Operations Presentation, January 2018.[26]

In retrospect, oil saved New England. Without the oil that ISO-NE required the natural gas power plants to keep on hand, New England would have suffered devastating blackouts.

Fuel Neutrality

In spite of ISO-NE success in using oil to save New England from blackouts, FERC later prohibited ISO-NE from using oil as an emergency fuel supply and forced ISO-NE to use auctions to keep the lights on. FERC, as a policy, insists on fuel neutrality for RTO/ISOs and disallowed the use of oil by ISO-NE for that reason.

ISO-NE modeled twenty-three scenarios for the winter of 2025 after the hard winter of 2018. Of these twenty-three scenarios, nineteen resulted in rolling blackouts. There is clearly a risk going forward.

The one-third of the country that does not belong to RTO/ISOs can do whatever makes sense, and use any fuel that provides low-cost, reliable service. Fuel neutrality for RTO/ISOs is an absurd piece of bureaucratic thinking that has no place in managing the grid. It's all about including wind and solar, and other renewables, to fight climate change. This is at the root of many problems inherent in RTO/ISOs.

As mentioned earlier, some members of FERC are attempting to force the elimination of fossil fuels. The paper published by Commissioner Glick starts out by saying: "The evidence that anthropogenic climate change is an existential threat to our way of life is incontrovertible."[17]

There are several factors adversely affecting ISO-NE:

1. FERC has now, in 2019, disallowed the use of oil storage to prevent blackouts.

2. CASPR is likely to result in more baseload power plants closing as they sell their contract obligations to renewable producers.

3. Certain baseload power plants are threatened with closure, making it even more difficult for ISO-NE to guarantee the availability of electricity and prevent blackouts.

   • Two nuclear power plants are threatened with closure: Millstone Nuclear Power Station, in Connecticut; and Seabrook Station, in New Hampshire. (Vermont Yankee shut down in 2014; Pilgrim Nuclear Power Station, in Massachusetts, shut down in 2019.)

   • Exelon Power announced the retirement of the Mystic Generating Station (natural gas), units 7, 8, 9, in Massachusetts; and the Mystic Jet (oil fueled peaking unit); combined rating of over 2,000 MW, by June 2022; but they will be kept on-line with out-of-market payments.

4. New York State has prevented the construction of new natural gas pipelines thereby preventing any new natural gas supply from the Marcellus shale to reach New England.

These factors, however, are not necessarily confined to ISO-NE, for example:

1. Freezing cold weather can also shut down wind power when temperatures get below 20 °F.

    • In 2019, "Wind generation in the Midcontinent Independent System Operator's [MISO] north region fell faster than anticipated during the polar vortex as turbines automatically shut down as temperatures fell below minus 20 degrees Fahrenheit."[28]

2. PV solar is susceptible to snow coverage and low insolation levels in northern areas, including in MISO, PJM, and NYISO.

3. Natural gas pipelines are under attack in many states where capacity expansions are needed.

4. Baseload plants are being closed in most RTO/ISOs.

The experience of ISO-NE during the winter of 2018 demonstrates how precariously the grid serving the people of New England teeters on failure, where failure means dangerous, life-threatening blackouts.

Chapter 17

The Failure of RTO/ISOs

The purported original purpose of RTO/ISOs was to improve the efficiency of competitive markets so as to lower the price of electricity to consumers while maintaining the reliability of the grid.

FERC Order 2000 supported the formation of RTO/ISOs:

> The Commission's goal is to promote efficiency in wholesale electricity markets and to ensure that electricity consumers pay the lowest price possible for reliable service.[29]

But after nearly twenty years, have RTO/ISOs achieved that objective?

If they have achieved that objective, residential prices for electricity in RTO/ISO markets should have decreased when compared with traditionally regulated markets. While prices for electricity have actually increased in all markets, the increases should have been less in states where the markets for electricity were under the control of RTO/ISOs. In fact, residential prices for electricity in the areas controlled by RTO/ISOs have increased more than in traditionally regulated markets.

Listed below are the seven RTO/ISOs and the traditionally regulated areas with the average price for electricity in 2003 and in 2017, together with the percent increases for each region and area.

At first blush, it would appear that RTO/ISOs have failed to deliver the promise of low prices for electricity. As a group, RTO/ISOs had a 64 percent increase in the residential price of electricity, while traditionally regulated markets had only a 53 percent increase. But two of the RTO/ISOs had only a 30 percent increase, so the average increase of the remaining RTO/ISOs was even greater than 64 percent.

Electricity Prices, cents per kWh

RTO/ISO	Avg. 2003	Avg. 2019	% Increase
ISO-NE	11.95	20.14	69%
NYISO	14.31	18.56	30%
PJM	8.05	12.89	60%
MISO	7.55	13.10	74%
SPP	7.38	12.26	66%
ERCOT	9.16	11.87	30%
CAISO	12.23	20.87	71%

		Average Increase	64%

Traditionally Regulated			
West	7.70	11.68	52%
South East	7.73	12.01	55%

		Average Increase	53%

Note: Regions as defined by FERC.

Table 3. Electricity prices.

The period from 2003 to 2018 should have been long enough to smooth out extraneous factors such as hurricanes in the southern states and polar vortexes in the northern states. The effect of low-cost natural gas after 2008 is fully accounted for by using this time frame. (See Appendix F.) RTO/ISOs have not achieved the goal established by FERC Order 2000, that "consumers pay the lowest price possible for reliable service."[30]

Some will argue that this is not a fair measurement of RTO/ISOs. For example, could an increase in population have required an inordinate amount of investment in RTO/ISO states? The fact is, population in five of the six RTO/ISOs have tended to shift to areas managed by traditionally regulated markets.

While there is no absolute measurement for whether reliability has decreased in RTO/ISO areas, there are indications it has. A report issued in January 2018 by ISO-NE Public, *Operational Fuel-Security Analysis,* warning that New England is likely to experience blackouts in the future.[31]

As mentioned in chapter 15 of this book, ERCOT (the RTO/ISO covering Texas) has seen its reserve margin become dangerously low, at 7.4 percent. In the past, its reserve margin was 13.75 percent. (See figure 8.)

The systems used by RTO/ISOs for establishing the price of electricity and ensuring adequate future supply have become so complex that complexity alone may lead to reduced reliability.

Performance

A closer look at the two RTO/ISOs that had only a 30 percent increase in the price of electricity is needed to help determine why they have performed better than the other RTO/ISOs.

NYISO

The NYISO had a disproportionate share of its electricity, i.e., 22 percent, generated by hydro. All the other RTO/ISOs had only very small percentages from hydro except CAISO, which had 12.3 percent.

The large amount of hydro helped stabilize prices. The new New York State mandates, established July 2019, for wind and solar, etc., will show whether the NYISO can continue with only modest price increases.

ERCOT

An examination of ERCOT data did not establish any obvious reason why ERCOT was able to keep price increases at bay. ERCOT only had the third highest amount of wind and solar in 2018, at 18.6 percent, while CAISO had 22.9 percent and Southwest Power Pool (SPP) had 24 percent.

ERCOT remains an anomaly.

Because RTO/ISO auction markets are rigged, they are driving base-load power off the grid. RTO/ISOs are built on a plethora of regulations, essentially a house of cards constantly in need of shoring up.

As Meredith Angwin says in her book, *Shorting the Grid:*

> As we have seen before, once you begin weaving a web of regulations, you need more regulations to fix things.[32]

The fact is, RTO/ISOs have failed to achieve the lowest prices possible for electricity, and are jeopardizing reliability.

While the jury may still be out as to whether RTO/ISOs should be abandoned, the evidence is crystal clear that traditionally regulated markets should never transition to RTO/ISOs.

Chapter 18

True Motivation

The original motivation for establishing RTO/ISOs in 1999 was probably to reduce costs while maintaining the grid's high level of reliability.

It would now appear that the motivation today is to cut CO_2 emissions by eliminating the use of fossil fuels and forcing the construction of wind and solar power generating facilities onto the grid. There is evidence that members of FERC and RTO/ISOs are forcing their views about climate change into the various procedures and systems controlling the grid.

Previous chapters have shown how auctions are rigged to allow wind and solar to beat out baseload competitors, forcing nuclear and coal-fired power plants off the grid.

The media places the blame on cheap natural gas, but the auction process is the means by which wind and solar are put ahead of baseload power plants. It's true that if wind and solar didn't exist, natural gas would win the bids, but that's irrelevant since subsidized wind and solar are a reality.

As of March 31, 2020, there are four commissioners on FERC, with Commissioner Danly added on March 31:

> Neil Chatterjee, Chairman
> Bernard L. McNamee
> Richard Glick
> James Danly

A review of FERC commissioners, and of published information from some RTO/ISOs, proves that auctions are the vehicle for forcing wind and solar onto the grid. Over time, the political process, where the Senate approves the appointment of commissioners, has resulted in the Commission being more or less equally divided between Republicans and Democrats. This has allowed politics, especially as it relates to climate change, to become a factor in FERC's rulings.

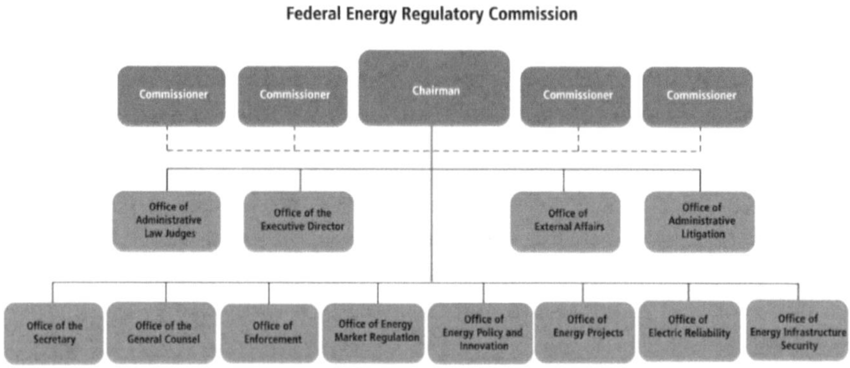

Federal Energy Regulatory Commission

Figure 14. FERC organization chart.

Chatterjee, McNamee, and Danly are Republican appointees, while Commissioner Richard Glick is a Democrat appointee. Cheryl LeFleur, a Democrat appointee, resigned in August 2019. (In September 2019, she was elected to the board of directors of ISO New England.)

The following quotations demonstrate how climate change has become a factor in FERC rulings.

- From RollCall, "LaFleur and Glick have pushed for more extensive reviews of the impact of natural gas pipelines and liquefied natural gas facilities on greenhouse gas emissions and climate change."[33]

- From Utility Dive, "With LaFleur on the commission, FERC was forced to reach a compromise to approve natural gas infrastructure opposed by Democrats for environmental reasons. . . . LaFleur's vote forced the inclusion of landmark greenhouse gas calculations in the approval order."[34]

- From a paper by the Energy Law Association, "FERC AND CLIMATE CHANGE," Rich Glick and Matthew Christiansen, Synopsis: The evidence that anthropogenic climate change is an existential threat to our way of life is incontrovertible. . . . the scope of its statutory responsibilities means that its decisions will inevitably affect the nation's greenhouse gas (GHG) emissions, and, therefore, climate change. As a result, the Commission is likely to become an increasingly important venue in the debate over how this nation will address climate change and those that want to address climate change will find that the Commission is an important agency with which to interact.[35]

- At a house hearing in June 2019, E&E news reported, "House Democrats questioned members of the Federal Energy Regulatory Commission yesterday over how they incorporate climate change considerations in their decision-making."[36]

It's clear that FERC's decisions have been affected, and will likely be even more affected in the future, by views about climate change. By what authority do FERC commissioners make judgments about climate change and impose those judgments onto all Americans? It is the EPA that is responsible for exercising authority over environmental issues such as climate change. Docket No. RM99-2-000; Order No. 2000, establishing the RTO/ISOs said:

> The Commission's goal is to promote efficiency in wholesale electricity markets and to ensure that electricity consumers pay the lowest price possible for reliable service.[37]

There is nothing in this docket that refers to climate change.

FERC's Mission

FERC's mission is as follows:

> FERC's Mission: Economically Efficient, Safe,
> Reliable, and Secure Energy for Consumers.
>
> Assist consumers in obtaining economically efficient,
> safe, reliable, and secure energy services at a reasonable
> cost through appropriate regulatory and market
> means, and collaborative efforts.[38]

There is nothing in FERC's mission or strategic plan that says it should address climate change or cut CO2 emissions.[39] FERC is not responsible for cutting oxides of nitrogen (NOx) or sulfur (SOx) or other emissions, and it should not be requiring the reduction of CO2 emissions. By surreptitiously forcing wind and solar onto the grid for the purpose of cutting CO2 emissions, FERC is arrogating the authority of the EPA. Climate change decisions made at FERC or at RTO/ISOs are harming Americans.

RTO/ISO Policy

RTO/ISOs have included climate change as a part of their operations. Proof of this is found in RTO/ISO documents.

A vice president of the PJM, RTO/ISO said,

> With new technology have come decreased emissions.
> Since the inception of the PJM market in 1997,
> emissions are at a 20-year low, . . .[40]

He made his interest in CO2 and climate change clear when he said CO2 had been reduced from 1,325 pounds per MWh in 2005 to 950 pounds per MWh in 2017. While he also referenced SO2 and NOx, he thought it important to include CO2.

He made his concern about climate change clear when he said,

> PJM markets have facilitated an unprecedented fuel and technology switch from older, less-efficient resources to advanced, increasingly efficient resources . . .[41]

That statement was biased and misleading because the most efficient methods for generating electricity are NGCC and nuclear power plants that provide baseload power. Presumably the PJM vice president was referring to natural gas power plants as the "older and less-efficient" resources, when they are, in fact, the most efficient and reliable resources on the grid. The capacity factors for NGCC and nuclear power plants are over 80 percent and 90 percent respectively, while capacity factors for the so-called "new and advanced technologies" are 34 percent for wind and around 20 percent for solar. In addition, NGCC plants have an efficiency of 63 percent while wind and solar are unreliable and intermittent.[42]

The PJM, RTO/ISO made it perfectly clear it was proposing to incorporate actions to combat climate change when it issued its paper in 2017, *Advancing Zero Emissions Objectives Through PJM's Energy Markets: A Review of Carbon-Pricing Frameworks.*

It's 2017 annual report said:

> PJM prefers a regional approach to carbon pricing in the energy market and believes a coordinated carbon policy could be advanced through the PJM markets . . . [43]

PJM is embroiled in the issue of carbon pricing and favors a common carbon price across its territory.

NYISO is more straightforward than PJM in its support of actions to address climate change. The 2019 annual *Grid and Market Report* features a graphic claiming a 51 percent reduction in CO2.

Quoting from the report:

> There is no historical precedent for the ambitious changes on the bulk power system envisioned by policymakers. . . . Policymakers seek even more aggressive

goals of 70 percent renewable energy by 2030 and 100 percent "clean energy sources" by 2040.[44]

And from NYISO's *Power Trend Report:*

> the NYISO is developing a proposal with stakeholders and policymakers to incorporate the societal costs associated with carbon dioxide emissions into its energy markets to better reflect the state's policy of reducing emissions.[45]

There can be no question that RTO/ISOs are acting to ensure that wind and solar drive fossil fuels off the grid. RTO/ISOs have taken it upon themselves to combat climate change. They are unelected bureaucrats deciding how other Americans are to live.

The RTO/ISOs operate behind the curtains, as did the Wizard of OZ, and it's extremely difficult to find out how they operate. There is very little transparency.

It can be argued that RTO/ISOs are merely reacting to pressure from state governments, but FERC is not responsible to local politicians and must not participate in anything beyond its mission. In addition, RTO/ISOs shouldn't be required to adjust their systems to compensate for state mandates. Or put more simply: RTO/ISOs shouldn't have to clean up the mess created by state politicians.

As it now stands, the higher costs and reduced reliability from wind and solar are hidden from the public.

Part Three

Nothing to Fear from CO2

The only thing we have to fear is . . . fear itself.

President Franklin Delano Roosevelt[46]

Chapter 19

History Is Important

Before considering the future effects of atmospheric CO2 on climate, it's important to remember how CO2 and natural forces have affected climate over the past ten thousand years.

This period, extending from the end of the last ice age, is known as the Holocene. A great deal is known about temperatures, CO2 levels, volcanic activity, and the sun's effect on the Earth during this period.

Figure 15 charts temperatures during the Holocene. We also know that atmospheric CO2 levels at the start of this period were 260 ppm, increasing to 280 ppm around four thousand years ago, and then remaining fairly constant at 280 ppm until the mid-1800s when they started to increase dramatically.

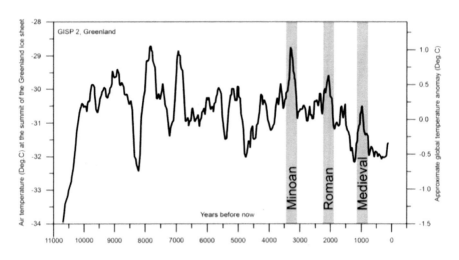

Figure 15. Holocene temperatures from Greenland Ice Sheet, reconstructed by Alley (2000) from GISP2 ice core data.

Since CO_2 levels had remained essentially constant until the mid-1800s, the only forces that could have affected temperatures during the Holocene before the mid-1800s were volcanic activity, the sun, and cosmic rays.[47] This is the most important conclusion to draw from figure 15.

It's obvious that temperatures varied during the Holocene. The data in figure 15 ends around 1850 and the latest warming is not included. While there is debate as to where we stand on the chart today, some have said that temperatures have been higher than today for most of the past ten thousand years. There is little doubt they were higher during the Medieval Warm Period, and certainly in the warm periods preceding it.

Volcanic activity generally cools the Earth and is reasonably well known, and is not responsible for the periods when temperatures were higher.[48] Thus, by deduction, it's fair to conclude that the sun was the main reason for higher temperatures over the past ten thousand years . . . at least until the mid-1800s.

These three forces, the sun, volcanic activity, and cosmic rays, didn't suddenly stop having an effect on temperatures when atmospheric CO_2 levels started to rise. Rising CO_2 levels are the new kid on the block, the new variable that must now be taken into consideration in addition to the sun, volcanoes, and cosmic rays.

The issue therefore is, how much do increasing levels of atmospheric CO_2 affect temperatures? The same thought process applies to methane and the other GHGs.

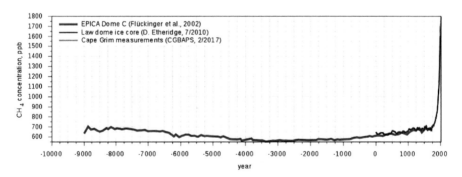

Figure 16. Methane in ppb.

Methane at the beginning of the Holocene (figure 16) was about 0.7 ppm, decreased to around 0.6 ppm, then increased gradually to 0.7 ppm by the mid-1800s, and then rose rapidly to around 1.88 ppm. Nitrous oxide levels were also relatively steady during the past ten thousand years until the mid-1800s.

Again, the only forces that could have affected temperatures until the mid-1800s were volcanic activity, the sun, and cosmic rays. These forces didn't suddenly stop exerting their effects on temperatures in the mid-1800s. They had a strong effect on temperatures then, and still do.

Chapter 20

Recent Trends

The Charney report, published in 1979, was an early attempt at trying to forecast how far temperatures would rise as the result of increased levels of atmospheric CO_2. It predicted that temperatures could rise by 6 °F with a doubling of CO_2 in the atmosphere.[49]

The Intergovernmental Panel on Climate Change (IPCC), Representative Concentration Pathway (RCP) 8.5 scenario in the 2014 AR5 report, predicted that temperatures could rise 8.6 °F by 2100.[50] While the RCP 8.5 was intended to show a worst case scenario, the media frequently referred to RCP 8.5 as "business as usual."

Then, in 2020, the Washington Post reported that RCP 8.5 was an unlikely extreme scenario. The article in the *Washington Post* said:

> Zeke Hausfather of the Breakthrough Institute, and Glen Peters, an energy expert at the Norwegian science organization CICERO, argue that the scenario ought to be discarded.
>
> Happily—and that's a word we climatologists rarely get to use—the world imagined in RCP 8.5 is one that, in our view, becomes increasingly implausible with every passing year.[51]

The RCP 8.5 has been discredited, which eliminates much of the doomsday hysteria found in the media.

Enough new scientific evidence has become available since the Charney report for people, who may have originally been alarmed by the report, to see that the climate is far less sensitive to CO_2 than originally thought. Actual temperature readings have shown that the IPCC computer projections have overstated the warming.

Dr. John Christy, Professor of Atmospheric Science, Alabama State Climatologist University of Alabama in Huntsville, in his Congressional testimony, used the following chart to demonstrate that the IPCC computer projections were overstating temperature rise. Figure 17 is from J. R. Christy's, 29 Mar 2017, Testimony before the US House Committee on Science, Space and Technology. Actual temperatures from balloons and satellites are far below the 102 computer model runs published by the IPCC.[52]

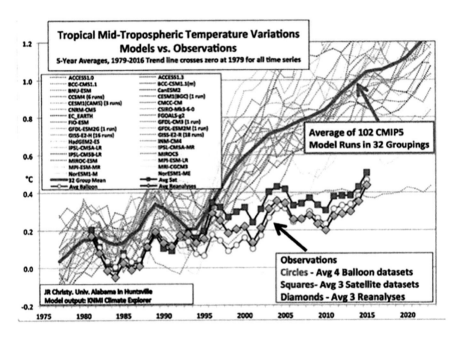

Figure 17. Tropical Mid-Tropospheric Temperature Variations.
J. R. Christy, 29 Mar 2017, House testimony.

The average of the 102 computer programs projected a temperature rise that was more than twice the actual temperature rise from readings taken by satellites and balloons. He also showed that, when IPCC computer models omitted GHG data, the program results came closer to corresponding to actual temperature readings, thus casting additional doubt on the IPCC projections (see figure 18.) In other words, when GHG data is omitted from the programs, the computer runs align with actual temperatures. Therefore, GHG data is distorting reality.

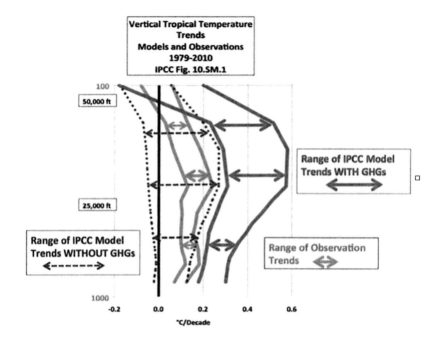

Figure 18. Vertical Tropical Temperature Trends.
J. R. Christy, 29 Mar 2017, House testimony.[53]

In 2017, Prof. Judith Curry said in her briefing "Climate models for the layman":

> There is growing evidence that climate models are running too hot and that climate sensitivity to carbon dioxide is at the lower end of the range provided by the IPCC.[54]

Then in 2020, Prof. Curry drew three main conclusions from her latest paper, "Plausible scenarios for climate change: 2020-2050":

1. We are starting to narrow the uncertainty in the amount of warming from emissions that we can expect out to 2050.

2. All three modes of natural variability–solar, volcanoes, internal variability–are expected to trend cool over the next three decades.

3. Depending on the relative magnitudes of emissions driven warming versus natural variability, decades with no warming, or even cooling, are more or less plausible.[55]

The accompanying table from Dr. Curry's paper shows the possibility of cooler temperatures.

	Warmest	Moderate	Coldest
Emissions	+0.70	+0.52	+0.35
Volcanoes	0	- 0.11	- 0.30
Solar	0	- 0.10	- 0.25
Oceans	0	- 0.20	- 0.30
NET	+0.70	+0.11	- 0.50

Table 4. From "Plausible scenarios for climate change: 2020–2050"
by Prof. Judith A. Curry.[56]

While the original Charney report projected a 6 °F rise for a doubling of CO2, and the IPCC projected as much as 8.6 °F warming by 2100, the latest science demonstrates that temperatures are forecast to rise much less than originally thought.

Dr. William Happer has provided additional scientific data that supports this contention. As can be seen from figure 19, temperature rise from a doubling of CO2 is barely noticeable. The top curve is the theoretical heat loss from the Earth into the vacuum of space for the range of frequencies, assuming no atmosphere. This is Planck's curve for heat loss from the Earth's blackbody. (Notations above the curves are of various chemical compounds at their spectral frequencies.)

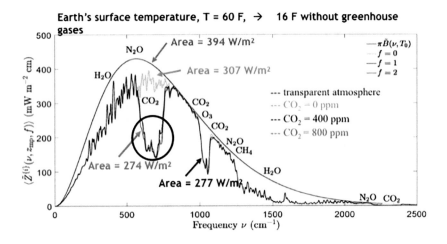

Figure 19. Graph from Dr. W. Happer's press briefing,
Madrid, Spain, December 2019.[57]

The sawtooth curve shows the actual heat loss through the Earth's atmosphere for each frequency, where the percentages of CO2 are 0 ppm, (in light-grey above the circle), 400 ppm (in black within the circle) and 800 ppm (in light-grey within the circle). The sawtooth curve is known as the Schwarzschild curve. (The heat loss for all other compounds are for conditions as they exist today.)

Of particular importance are the circled, black, and light-grey CO2 curves. These two curves, highlighted by the circle, are virtually the same, indicating that heat loss is nearly unchanged after doubling CO2 from 400 to 800 ppm. With respect to methane, Dr. Happer displayed a second curve showing that a doubling of methane would also have little effect on temperatures.

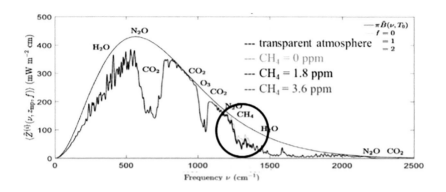

*Figure 20. Graph for CH4 from W. Happer's press briefing,
Madrid, Spain, December 2019.*[58]

The sawtooth curve shows the actual heat loss through the Earth's atmosphere for each frequency, where the percentages of CH4 (methane) are 0 ppm, (in light-grey sawtooth above the dark-grey curves), 1.8 ppm (in black) and 3.6 ppm (in light-grey). Again, the black and light-grey curves are virtually on top of each other, showing that the heat loss is nearly unchanged with a doubling of methane.

The fear has existed that higher global temperatures would release overwhelming amounts of methane from the permafrost, or tundra, but the latest scientific study has eliminated that possible threat. The new study from the Scripps Institution of Oceanography and the University of Rochester establishes that there won't be an uncontrollable release of methane from the tundra because of warming. The study published in the journal *Science* found that tundra, i.e., permafrost, that "melted during a previous warm period between 18,000 and 8,000 years ago released little methane."[59]

The scientists drilled ice cores from Taylor Glacier in Antarctica and found trapped air which could be analyzed to determine the age of any methane found in the atmosphere. The scientists said:

> We can say with great confidence now, that tundra methane did not contribute to atmospheric methane at all, during the last time that the world warmed a lot.[60]

This unlikely threat has been eliminated, and we can rely on Dr. Happer's curve to show there will be very little temperature increase from a doubling of methane. The latest science has established we have nothing to fear from increases in atmospheric levels of CO2 or methane. Drastic actions to cut CO2 and methane emissions are not warranted, and such actions will have destructive effects on the economy and our way of life.

We can see from the historic record that the sun has always had a major influence on the Earth's climate, and recent temperature trends have established that temperatures are likely to be cooler than projected in the media, no matter what the cause.

We should not be taking actions out of fear to cut CO2 and methane emissions. There is every reason to believe we are not facing an existential threat, and that actions to cut CO2 and methane will cause severe harm to Americans . . . and mankind. Dozens and dozens of scientists are not fearful of GHG being an existential threat. (Appendix G lists some of these scientists.)

Chapter 21

Abandoning Fossil Fuels

There is a massive effort underway to convince people they need to destroy their lifestyles in order to save the world from a climate catastrophe. There are two fundamental reasons why it is profoundly wrong to abandon fossil fuels and implement policies to install wind and solar generation, as some of the RTO/ISOs and state governments have been doing.

First, CO_2 is, in all likelihood, not an existential threat to mankind. The preceding chapters explain in some detail why GHGs are not an existential threat to mankind and why the sun is probably the primary cause of climate change.

There is strong evidence, dating back to William Herschel in the 1700s, that the sun has always played an important role in the Earth's climate. The Carrington Event in 1859, a huge solar flare, demonstrated that the sun is directly linked to the Earth. The sun can reach out and touch the Earth.

A solar flare today of the same magnitude as the Carrington Event, in 1859, could threaten the survival of millions of people. For the past ten thousand years, up until the mid-1800s, it was only the sun, with occasional volcanoes, that affected the Earth's climate. And the sun didn't stop having an effect on the Earth's climate just because mankind started to use fossil fuels. Scientists such as Nir Shaviv, H. Svensmark, and Willie Soon have proposed hypotheses for how the sun interacts with the Earth to affect its climate.

We are also being subjected to an effort to change our economic system from capitalism to socialism. Christiana Figueres, executive secretary of the United Nations Framework Convention on Climate Change (UNFCCC), made this very clear when she said,

This is probably the most difficult task we have ever given ourselves, which is to intentionally transform the economic development model, for the first time in human history. This is the first time in the history of mankind that we are setting ourselves the task of intentionally, within a defined period of time, to change the economic development model that has been reigning for the, at least, 150 years, since the industrial revolution.[61]

The world's economic model is based on capitalism.

On the face of it, Figueres made it clear it will be necessary to do away with capitalism to achieve the objective of cutting CO2 emissions. The United States is a signatory to the UNFCCC treaty and the Senate has ratified the treaty so the United States is obligated to comply with the treaty.

And Naomi Klein said, "Forget everything you think you know about global warming. It's not about carbon—it's about capitalism."[62]

Earlier in 2010, Ottmar Edenhofer, economist and co-chair of the IPCC Working Group III, said,

Climate policy has almost nothing to do anymore with environmental protection, the next world climate summit in Cancun is actually an economy summit during which the distribution of the world's resources will be negotiated.[63]

Some of the impetus for abandoning fossil fuels comes from socialists who want to promote the climate change issue to their advantage. This emphasis on abandoning fossil fuels merely increases the risk to grid reliability.

Second, it's impossible to eliminate the use of fossil fuels without destroying modern living standards. In 2020, we suffered through the terrible COVID-19 virus that killed thousands, and only fossil fuels can provide the equipment and resources to combat such viruses. Medicines and plastics are made from oil and natural gas. Steel is required in the manufacture of medical equipment, and steel requires natural gas or hydrogen. Ventilators, which were in such short supply during the COVID-19 pandemic, require plastics made from oil and natural gas. Ventilators also require a reliable supply of electricity. Vaccines are injected using steel needles mounted on a plastic or glass syringe body. And like cement, glass requires very high temperatures produced from natural gas, oil, or coal for its manufacture.

Over 80 percent of the energy used by the world is derived from fossil fuels. Figure 21 shows that the preponderance of fuels used by mankind are natural gas (methane), oil, and coal.

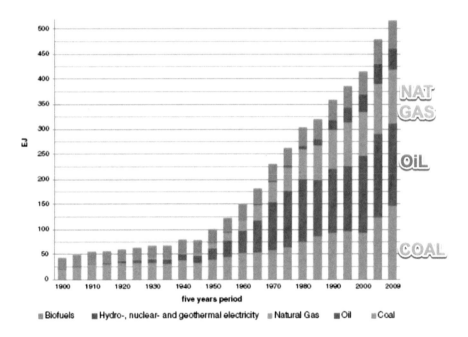

Figure 21. Types of fuels used by world.
Source: Bithas and Kalimeris, 2016[64]

Over the last twenty years, billions of dollars have been spent on wind, solar and biofuels, but over 80 percent of the world's energy still comes from fossil fuels. While there are many fanciful ideas about how to eliminate the use of fossil fuels, all are more costly, and few of them are technically feasible.

One simple example illustrates the point. Cement and steel can't be made without emitting CO_2, unless hydrogen is used in combination with CCS. CCS is needed because limestone emits CO_2 during the cement-making process.

Hydrogen would have to be produced using electrolysis because the traditional process for producing hydrogen uses methane, which emits CO_2. The proposal to replace natural gas with hydrogen is a costly delusion being pursued in Europe. Appendix C describes Europe's nascent attempt to transition to hydrogen and its use for making steel and cement. It explains the many obstacles to a hydrogen economy.

Someday, fusion energy will become available, but we have been predicting that for the past seventy years. We should not accept that 5.8 billion people in the world who live in undeveloped or developing countries, such as in sub-Saharan Africa, should be forced to live in squalor and poverty.

The facts are clear. There is no need to eliminate the use of fossil fuels, and it's unethical and virtually impossible to do so anyway.

Part Four

Last Word

Acta non Verba, Deeds not Words.
Motto, United States Merchant Marine Academy

Chapter 22

The War on Fossil Fuels

The war on fossil fuels is a moral issue as well as an economic and energy issue.

Efforts to eliminate the use of fossil fuels don't only affect Americans, they have a huge implication for the poorest of the poor around the world. There is an unfolding tragedy as fossil fuels are banned or economically black balled. The World Bank, for example, has prohibited financing for building coal-fired power plants. It has also stopped financing oil and natural gas development.[65]

The people of Africa are suffering from the war on fossil fuels.

The war on fossil fuels is literally killing people. Here is how McKinsey & Company describes the current situation.[66]

> There is a direct correlation between economic growth and electricity supply. If sub-Saharan Africa is to fulfill its promise, it needs power—and lots of it. Sub-Saharan Africa is starved for electricity.
>
> It has 13 percent of the world's population, but 48 percent of the share of the global population without access to electricity.

Here is a comparison that should shock any American.

While the average American consumes 12,000 kWh/year of electricity, average consumption in the Central African Republic is 33 kWh/year, and in Chad it's only 13 kWh/year.[67] This is appalling, and tragic.

Sub-Saharan Africa is an area beset with poverty and short life expectancy. Each person in Chad has enough electricity to use one

40- watt light bulb for one hour a day. And, most people in Chad can't use any light bulbs because electricity isn't available at their homes.

Pictures taken from outer space, such as the one shown in figure 22 of Africa at night, show a continent in dire need of electricity—electricity that could be generated using fossil fuels.

Figure 22. NASA Satellite image of Africa at night.

Water and modern cooking are perhaps two of the most important advances needed in sub-Saharan Africa. Water is needed for sanitation. It's been estimated that 10 percent of children under five years of age in sub-Saharan Africa die from diarrhea.[68]

Pumps are needed for pumping water from wells. A ½-HP motor running thirty minutes a day would use 240 kWh per year, yet wells are in short supply in sub-Saharan Africa. This is only one of the many tragedies created when there is a lack of electricity. Too many people are still using dung and wood to cook their meals.

Electricity would go a long way to solving both problems.

Three resources are available in sub-Saharan Africa that could be used to generate electricity: coal, natural gas, and oil. All are fossil fuels. Hydro and geothermal are also available in some areas and should be used when available.

While there are other problems in Africa, such as graft and religious warfare, the cheapest and most easily used resource is likely unavailable for building power plants in sub-Saharan Africa because of the war on fossil fuels.

The 300 MW Lake Turkana wind farm in Kenya that cost around $800 million has been highly touted for being renewable.[68]

Assuming a CF of 35 percent for the wind farm, which is located in an area that reportedly has excellent wind, a modern HELE coal-fired power plant with all the appropriate pollution controls could have been built that would provide twice as much reliable electricity from the same investment.

In addition, the wind turbines will have to be replaced in around twenty years, while the coal-fired power plant could remain in operation for sixty years. Finally, the wind farm will be unreliable, with wind speeds varying throughout the day and night, and probably stopping intermittently when wind speeds drop below six mph.

The Lake Turkana wind farm cost hundreds of millions of hard to-come-by dollars, much of it wasted because they could have built a reliable coal-fired power plant in its place. These wasted dollars are dollars that poor countries desperately need.

Tanzania could have done even better if it had built a NGCC power plant, though natural gas itself, though available, requires more development.

While the situation in Africa is hard to comprehend, other countries are also harmed by the war on fossil fuels. India and Indonesia, two very important and populous countries, need electricity.

In India, the average person consumes only 824 kWh/year, while in Indonesia, it's only 780 kWh/year. While this is substantially greater than in sub-Saharan Africa, it is still only a fraction of per capita consumption of electricity in developed economies.

Indonesia has large coal reserves, for coal-fired power plants and export, to sustain its economy. India has large coal reserves that it is trying to develop, primarily so it can generate more electricity. Should we expect these countries, with large populations, to relegate their citizens to continued poverty because of the war on fossil fuels?

Irrigation of rice paddies and farms require water distributed to them by electricity or fossil fuel driven pumps. There are foot driven pumps, using a man walking on them, but that in itself is a tragic waste.

The war on fossil fuels is a tragedy because it's condemning people to poverty and an early death.

We should be encouraging economic development not discouraging it. Yes, battery powered cars may be preferred in rich economies, but gasoline-powered scooters and three-wheel vehicles are cheaper and better suited for a developing country.

The use of fossil fuels is beneficial to mankind, and the war on fossil fuels is immoral.

Mankind isn't addicted to fossil fuels. Instead, mankind needs fossil fuels to emerge from poverty and then sustain a healthy lifestyle by which children can grow to become productive adults.

Chapter 23

Conclusion

FERC, the regulatory body that is supposed to protect Americans by guaranteeing the lowest possible price for electricity and ensuring a grid that's able to reliably deliver electricity to all Americans, has become politicized over the climate change issue.

FERC is forcing its views on climate change into the procedures and systems that determine the price and availability of electricity, and the reliability of the grid. FERC, for example, by adhering to a rigid fuel neutral policy, has prevented the ISO-NE from using the best fuel to ensure grid reliability and guarantee the safety of people living in New England. It has arrogated the EPA's responsibilities.

RTO/ISOs have evolved into a shell game whereby ordinary citizens and families are paying more for electricity. Wind and solar installations receive subsidies paid for by taxpayers and then also receive payments from day-ahead and real-time auctions and possibly from capacity auctions, ultimately paid for by the same taxpayers in their role as consumers.

If there is a constant increase in wind and solar installations, then natural gas, nuclear, and coal-fired power plants will be driven off the grid to the detriment of all Americans.

The workarounds, such as MOPR and CASPR, and out-of-market efforts to maintain nuclear power, are not working. Special interest groups are forcing fossil fuels off the grid, and these out-of-market efforts should be stopped. They are not necessary, and they are destructive.

The uniform clearing price auction system is rigging auctions in favor of wind and solar, as explained in chapters 10 and 11, and is driving low-cost electricity off the grid while also lowering grid reliability.

It's essential that the uniform clearing price auction be replaced with auctions where bidders get paid what they bid.

RTO/ISOs need to be investigated to determine whether they should be abandoned in favor of traditionally regulated markets. Chapter 17 provides data demonstrating that traditionally regulated markets have performed better overall than RTO/ISOs in providing low-cost and reliable electricity. The idea that RTO/ISOs should be abandoned and that all areas of the country should return to a regulated market is not a novel idea. A paper by Wilkinson, Barker, Knauer LLP, in 2018, said:

> If regulatory and legal obstacles continue to foreclose these legislative or administrative actions, then the only remaining option is to vertically reintegrate.[70]

Vertically integrated regulated markets are the fairest way to manage the generation and distribution of electricity. It's also the best way to ensure grid reliability. Vertically integrated regulated markets will ensure that the method chosen for generating electricity will be based on the lowest cost and most reliable method.

Reliability is crucial. We now know pandemics are a real threat. We must maintain reserve margins that ensure adequate supplies of electricity to meet any future threat. We cannot allow the reliability of the grid to be undermined through faulty thinking and poor planning. We must use fuels that can supply baseload, reliable power, and preferably, those sources that generate electricity at the lowest cost.

Pandemics seem to strike during the winter, and winter is when the grid is most vulnerable to the effects of snow, ice, freezing rain, and cloudy weather. These are the conditions that are most likely to negatively affect wind turbines and solar arrays.

To guarantee the availability of electricity for any threat, only baseload power—using natural gas, coal-fired, or nuclear power plants—can provide reliable supplies of electricity, day and night.

Appendix A

Acronyms

AAR	Auction Revenue Right
CAES	Compressed Air Energy Storage
CAISO	California Independent System Operator
CASPR	Competitive Auctions with Sponsored Policy Resources
CCS	Carbon Capture and Sequestration
CH4	Methane
CO2	Carbon Dioxide
CONE	Cost of New Entry
DR	Demand Response
EIA	Energy Information Administration
ELCC	Effective Load Carrying Capability
EPA	Environmental Protection Agency
ERCOT	Electric Reliability Council of Texas
FERC	Federal Energy Regulatory Commission
FIT	Feed In Tariffs
FTR	Financial Transmission Rights
GHG	Greenhouse Gas
IPCC	Intergovernmental Panel on Climate Change
ISO	Independent System Operator
ISO-NE	Independent System Operator – New England
KW	Kilowatt = 1,000 watts
LCOE	Levelized Cost of Electricity or Energy
LOLE	Loss of Load Expectation
MISO	Midcontinent Independent System Operator
MOPR	Minimum Offer Price Rule
MW	Megawatt = 1,000,000 watts
MWh	Megawatt hour

NERC	North American Electric Reliability Corporation
NGCC	Natural Gas Combined Cycle power plant (Also referred to as CCGT or CCPP)
NREL	National Renewable Energy Laboratory
NYISO	New York Independent System Operator
OATT	Open Access Transmission Tariff (OATT)
PJM	PJM is an RTO covering the states of Delaware, Illinois, Indiana, Kentucky, Maryland, Michigan, New Jersey, North Carolina, Ohio, Pennsylvania, Tennessee, Virginia, West Virginia, and the District of Columbia. (In 1956, two companies joined and renamed as "Pennsylvania-New Jersey-Maryland Interconnection," or PJM.)
PPA	Power Purchase Agreement
PV	PhotoVoltaic
RCP	Representative Concentration Pathway
REC	Renewable Energy Credit
RMR	Reliability Must Run
RPM	Revolutions Per Minute
RPS	Renewable Portfolio Standards
RTO	Regional Transmission Operator
RTR	Renewable Technology Resource
SMR	Steam Methane Reforming
SO2	Sulfur Dioxide
SPP	Southwest Power Pool
WAPA	Western Area Power Administration

Appendix B

Decline of Nuclear Power

in the United States

There are currently ninety-eight nuclear power plants operating in the United States. These ninety-eight plants produce 19 percent of the electricity in the United States.

However, five U.S. nuclear plants closed between 2015 and 2020, and at least five more are scheduled for closure by 2025. An additional group of nuclear plants will begin closing in 2029 as their operating licenses expire.

Nuclear power plants were given an initial forty-year operating license, and most of the plants, at least eighty-seven, have received a twenty-year extension to their operating license. At least three nuclear power plants have received a second renewal allowing them to operate for an additional twenty-years, or for a total of eighty years. Another fifteen or so plants are likely to apply for a second renewal. But, even with a second renewal for some plants, every existing nuclear power plant in the United States will be shut down by 2064.

Four more nuclear power plants were under construction in 2013, but the contractor, Westinghouse, went bankrupt. As a result, the two units being built in South Carolina were canceled, while construction of the Georgia Power's Plant Vogtle units 3 and 4 has resumed under an economic cloud.

Existing nuclear power plants are closing for two reasons:

1. Auctions have allowed subsidized wind and solar power plants to enter very low bids, as described in chapter 9, while low natural gas prices has allowed NGCC power

plants to also bid below nuclear power plants. This has forced nuclear power plants to close.

2. Unwarranted fear of radiation has frequently created a political environment favoring the closure of nuclear power plants. The most recent example is the closing of the units at Indian Point, located thirty-six miles north of New York City, on the Hudson River.

Fear of Radiation

Some organizations, such as the Natural Resource Defense Council (NRDC) and Greenpeace, have opposed nuclear power because of nuclear proliferation and fear of radiation.

With around seventy new nuclear power plants being built around the world, it would seem as though any possibility of stopping the proliferation of nuclear weapons has long since passed.

<u>Three Mile Island</u>

A cooling malfunction caused part of the core to melt in the #2 reactor. Virtually no radiation escaped from the island.

Fear was created by the movie *China Syndrome*, when the reporter, played by Jane Fonda, was told that a meltdown would send a cloud of radiation covering the entire state of Pennsylvania. This was a lie that Fonda did not dispute or investigate. It is what the title of the movie was meant to portray.

It's impossible for a melted reactor core to burrow into the ground and reach the water table. Every reactor in the United States, and probably other countries, has a thick concrete, steel-reinforced base under the reactor, where any melting reactor material would be caught.

The accident at Three Mile Island resulted in no injuries, adverse health affects, or deaths from radiation.

Chernobyl

Chernobyl was no accident. It was the result of incompetence and negligence.

The reactor design was prone to being unstable at low power, yet an unauthorized test was conducted under low-load conditions that led to the meltdown and subsequent graphite fire.

No modern reactor in the world has conditions similar to those that created the Chernobyl disaster, where there was also no containment vessel.

The first responders who were attempting to extinguish the graphite fire were the only people who died from radiation. The United Nations Scientific Committee on the Effects of Atomic Radiation (UNSCEAR) reported fifty-five deaths from radiation.

Fukushima Daiichi

The Fukushima disaster was caused by a tsunami.

The four nuclear power plants shut down properly after the earthquake that initiated the events. A tsunami flooded out the diesel generators that provided backup power after the connection to the grid was lost. Backup power was needed to maintain the flow of cooling water throughout the plants. All four reactor cores melted down and were kept within the building.

There were no deaths from radiation.

Generation 3 Reactors

New reactors, including the units being built in Georgia, are designed to provide a supply of cooling water without the need for diesel backup generators. This eliminates a source of concern.

New Reactors

The cost of building new reactors, similar to those being built in Georgia, has ballooned out of control. These reactors now cost more than $6,000 per KW to build in the United States.

For comparison: A new HELE coal-fired plant would cost $2,800 per KW, while a new wind turbine would cost $2,000 per KW, and a new NGCC power plant would cost $1,000 per KW.

The future of nuclear power is likely to rest with smaller units, of modular design, or with ongoing research of other potential designs.

Modular reactor designs have been proposed by several companies, and some designs are being evaluated by the Nuclear Regulatory Commission.

Some of these units are small enough to be buried underground. All are designed so that an initial small unit can be added to, module by module.

Storage of Nuclear Waste

The storage of spent reactor fuels has been a continuing concern. Currently, spent material is being stored in pools or in steel casks at the nuclear power plants.

Long term storage was to be done at a special facility. Yucca Mountain, Nevada, was chosen for that purpose, but political pressure has resulted in postponement, if not abandonment, of the site.

With Yucca Mountain storage seemingly going nowhere, and with the public emotionally opposed to nuclear power, and all existing plants closing by 2064, it would seem that nuclear power is in terminal decline in the United States.

Appendix C

Hydrogen

Europe was one of the first to propose using hydrogen to allow it to achieve its climate goals.

Hydrogen is rarely found in a free state and is virtually always combined with some other element. The most common compound is water, H2O. Nearly all hydrogen has been made from natural gas, CH4, using methane reforming, which results in the release of CO2. An alternative method for producing hydrogen is to use electrolysis, which requires large quantities of electricity. Since methane gas is an anathema to those who believe there is a catastrophic threat from climate change, they insist on using electrolysis with electricity sourced from renewables.

This gives rise to a tier of hydrogens:

- grey hydrogen – made from methane using steam methane reforming (SMR)

- brown hydrogen – made from coal gas using SMR

- blue hydrogen – made from methane gas using methane reforming with CO2 capture and sequestration

- green hydrogen – produced from water using electrolysis where the electricity is generated by renewables, such as wind and solar

The paper, "50% Hydrogen for Europe: a manifesto," outlines a plan for Europe to produce hydrogen and eliminate the use of natural gas. The green hydrogen would be produced by electrolysis using wind and solar to generate the needed electricity. There are three problems with this fantasy: (1) the existing gas distribution system, (2) sufficient wind and solar capacity, and (3) building the hydrogen pipeline.[71]

Existing System

First there is a serious question about using hydrogen in the existing gas distribution system throughout Europe.

The National Renewable Energy Laboratory (NREL) points out in its paper, "Blending Hydrogen into Natural Gas Pipeline Networks: A Review of Key Issues," that natural gas pipelines can only safely carry a mixture containing less than 20 percent hydrogen.[72]

Another restriction on using hydrogen in existing gas lines is that some appliances, such as gas stoves, can't safely use gas with levels of hydrogen higher than 20 percent.

Another important complication is that the gas distribution system contains a variety of pipes—including copper, iron, and plastics—with each material reacting differently to hydrogen.

Hydrogen concentrations higher than 20 percent can cause embrittlement in copper piping resulting in cracks and leakage, while hydrogen in plastic piping can leak through the plastic pipe's walls.

The European system is similar to the natural gas pipeline system in the United States, and in fact, may be even more complicated since much of the gas distribution system was built prior to the formation of the EU with different countries having differing regulations.

The paper, "50% Hydrogen for Europe: a manifesto," makes the following claim:

> The existing gas transmission and distribution infrastructure is suitable for hydrogen with minimal or no modifications.[73]

Such a claim flies in the face of NREL's studies. Expediency could easily sacrifice safety.

Wind and Solar Capacity

The second problem with transitioning to 100 percent green hydrogen is having sufficient wind and solar capacity to generate enough electricity to produce the required quantities of hydrogen after building all the wind and solar farms to satisfy the need for electricity for everyday residential and industrial use, and for battery powered vehicles.

No one has addressed this question, but Germany has reduced its CO_2 emissions by only 31 percent and is having difficulty building sufficient wind and solar farms to meet the objectives of its Energiewende program, let alone building more wind and solar for a hydrogen program.

Hydrogen Pipeline

But, even if these two problems can be overcome, building the hydrogen pipeline backbone is a huge undertaking.

The hydrogen backbone pipeline, figure 22, is approximately 7,000 miles long with some of it under water. There have been less than 2,000 miles of hydrogen pipeline built over the past decades in Europe, so a great deal of learning experience will be added to any cost estimates.

Referencing an *Oil & Gas Journal,* 2016 report, the average cost of natural gas pipelines built in the United States that year was $7.5 million per mile.[74] (Due to the lack of cost history of building hydrogen pipelines, a second paper is referenced that explores the various material, labor and miscellaneous costs. The most important cost is apparently related to whether pipelines are constructed in populated areas, and Europe has greater population density than the United States.[75])

This would suggest that the cost of a 7,000 mile hydrogen pipeline will be at least $52 billion. Since hydrogen pipelines will require special materials, joints, and compressors to prevent leakage, the cost is very likely to be at least this high.

Steel and cement cannot be made without fossil fuels except by replacing coal and natural gas with hydrogen. The making of steel and cement produces large quantities of CO_2.

Figure 23. Facsimile of map from "50% Hydrogen for Europe: a manifesto" illustrates approximate location of 7,000 mile natural gas hydrogen backbone.

Steel

The steel industry accounts for 7–9 percent of worldwide CO_2 emissions,[76] while cement production accounts for 5 percent.[77]

Salzgitter, a large German steel producer, is developing a "technically feasible but not economically viable" concept to replace fossil fuels, mostly coal used in blast furnaces, with green hydrogen.[78] But Salzgitter isn't proceeding with the replacement of its blast furnaces because it can't do so without government intervention. It requires $1.5 billion in government subsidies just to begin making the required investment, and the steel that's produced is far more expensive than steel made using coal.

In other words, Salzgitter must invest large sums in processes that are more expensive than existing processes, which results in making steel that can't be sold in a competitive market.

Cement

When making cement, much of the CO2 is from the material itself, while some of it comes from the fuel needed to create the heat that's required to convert limestone to cement. The article "Emissions from the Cement Industry" explains that because of the extreme heat required, "Cement manufacturing is highly energy- and emissions-intensive":

> The primary component of cement is limestone. To produce cement, limestone and other clay-like materials are heated in a kiln at 1400°C and then ground to form a lumpy, solid substance called clinker; which is then combined with gypsum to form cement.[79]

Heating limestone releases CO2 directly, while the burning of fossil fuels to heat the kiln also results in CO2 emissions. Roughly 50 percent of the overall emissions of CO2 from the making of cement are from the limestone, and these emissions are nearly impossible to eliminate.

Hydrogen Fantasy

Green hydrogen, assuming it can be safely used in pipelines, doesn't address some of the other mundane problems associated with hydrogen. Transporting hydrogen as a liquid, which will be needed where pipelines aren't available, incurs large energy losses.

The cost of hydrogen produced from electricity generated by wind and solar will be more expensive than the fuels used today, such as gasoline and natural gas, thereby burdening a hydrogen economy with additional added costs.

The proponents of climate change are tossing hydrogen into the energy mix as a Hail Mary pass in an attempt to salvage their fantasy of eliminating fossil fuels.

Appendix D

Cost of Electricity

For years, the levelized cost of electricity (LCOE) was viewed as an accurate method for comparing the relative costs for each method of generating electricity. In recent years, different entities, including the EIA have put forth new methods for calculating LCOEs. The EIA no longer reports actual LCOEs, but merely estimated LCOEs for sometime in the future, such as for 2024.

Lazard was widely reported to have produced LCOEs that showed wind and solar as having LCOEs below or competitive with NGCC, coal-fired, and nuclear power plants.[80] Chapter 9-3, *Energy: The Source of Prosperity*,[81] describes in detail how LCOEs have been manipulated. Lazard, for example, used "undefined resource availability" in place of published, and agreed upon, insolation levels, and published tables of wind speeds at various elevations. The concept of undefined resource availability is opaque; it's best to read the text from *Energy: The Source of Prosperity* for a clearer understanding of the term.

Energy: The Source of Prosperity (chapter 9-3), makes the following conclusion:

<u>Conclusion</u>

If an undefined "resource availability" is used to calculate LCOEs, the resulting LCOEs can't be compared with a traditionally derived levelized cost of electricity (LCOE): It's like comparing cashews with apples.[82]

Cost of Electricity

Pawning these LCOEs off as equivalent to traditionally calculated LCOEs is misleading at best, and at worst, could be considered deceptive.

In addition, wind and solar are unreliable, and LCOEs do not reflect the extra costs associated with having to compensate for their intermittency and unreliability. Wind and solar need backup, and backup costs money.

The Lazard report and virtually all media articles attempting to compare LCOEs between wind and solar and coal and natural gas are flawed and meaningless. Wind and solar cannot replace coal and natural gas on a one for one basis—they are not interchangeable LEGO pieces.

Coal-fired and natural gas combined cycle power plants, along with nuclear power plants, continue to be the least costly methods for generating electricity, notwithstanding the latest study from biased observers.

Appendix E

Credit Multipliers

The following is a quote from *Clean Energy State Alliance*:

> The most popular mechanisms for targeting specific technologies or applications are RPS carve-outs and credit multipliers. A carveout serves as a subset of a larger RPS, requiring a certain percentage of the overall requirement to be met with a specific technology or application. Credit multipliers, on the other hand, award more than one (or less than one) renewable energy certificate (REC) for electricity produced by certain technologies or applications.[83]

Note that the use of renewable energy certificates (REC) can also increase the price paid by consumers for electricity. The utility buying RECs increases its costs, which must eventually be recovered from higher prices for electricity.

Credit multipliers haven't been used recently.

Appendix F

Natural Gas Historic Pricing

Prior to around 2008, it was believed the United States was running out of natural gas and that natural gas would have to be imported from the Mideast or Australia. Import terminals were being built in the United States to allow the importation of liquified natural gas. The price of natural gas peaked at around $13 per million BTU.

As the result of fracking, natural gas supplies increased dramatically and the price of natural gas plummeted.

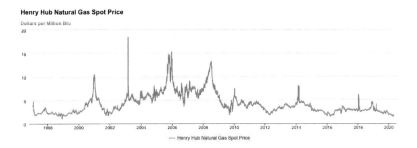

Figure 24. Henry Hub Natural Gas Prices, $ per million BTU.[84]

It has averaged around $3.00 per Million BTU since 2015 and is currently below $2.00 per Million BTU.

Appendix G

Scientists

There are dozens of climate scientists who have, at one time or another, said that CO2 is not an existential threat to mankind or that mitigation and adaptation are the best responses to climate change.

In addition, there are dozens of other highly qualified scientists of various disciplines who have participated in various climate studies, and a few of them are also listed below. There are another 32,000 engineers and scientists who have signed a petition stating, in part:

> There is no convincing scientific evidence that human release of carbon dioxide, methane, or other greenhouse gases is causing or will, in the foreseeable future, cause catastrophic heating of the Earth's atmosphere and disruption of the Earth's climate. Moreover, there is substantial scientific evidence that increases in atmospheric carbon dioxide produce many beneficial effects upon the natural plant and animal environments of the Earth.[85]

There is a political website, DeSmogblog.com, that may have accused many of those on the following lists of being climate change skeptics or deniers. Interestingly, those publishing DeSmog consist of three reporters and an environmental lawyer, none of whom, according to the DeSmog website, have engineering or science degrees. DeSmog pits reporters against highly educated scientists.

Climate Scientists

The following is a list of only a few scientists who have said CO2 is not an existential threat to mankind or that mitigation and adaptation are the best responses to climate change. The list includes a Nobel Laureate and IPCC reviewers.

Judith Curry, PhD
William Happer, PhD
John Christy, PhD
Roy Spencer, PhD
Richard Lindzen, PhD
Ivar Giaever. PhD
Roger A. Pielke Sr., PhD
Sallie Baliunas, PhD
Timothy F. Ball, PhD
Patrick J. Michaels, PhD
Robert M. Carter, PhD
Dr. C. Schlüchter, PhD
Dr. Michael Schnell, PhD
H. Abdussamatov, PhD
Ian Plimer, PhD
Syun-Ichi Akasofu, PhD
Dr. Peter Ridd, PhD
Claude J. Allègre, PhD
Sebastian Lüning, PhD
Freeman Dyson, FRS
Willie Soon, PhD

Nir J. Shaviv, PhD
Vincent Gray, PhD
William Gray, PhD
Howard Hayden, PhD
David Legates, PhD
Anthony R. Lupo, PhD
James O'Brien, PhD
Andreas Prokoph, PhD
Laurence I. Gould, PhD
Nicola Scafetta, PhD
Boris Winterhalter, PhD
Dr. Lutz Niemann, PhD
Don J. Easterbrook, PhD
John S. Theon, PhD
Thomas P. Sheahen, PhD
Ralph B. Alexander, PhD
Petr Chylek, PhD
Michael R. Fox, PhD
Chris de Freitas, PhD
Henrik Svensmark, PhD

Multidisciplinary List

The following is a list of scientists from multiple disciplines who have participated in some way with the climate change issue, either demonstrating that the effects of GHG are not existential threats or establishing the historic effects of CO2 during and before the Holocene Epoch. There are no lawyers or economists on this list though they have, on occasion, been involved with the climate change issue.

At the end of the list is the group of NASA scientists and engineers who put a man on the moon. These Apollo scientists were able to conquer space and certainly had the scientific background and wherewithal to address climate change.

Dr. Susan Crockford, PhD	Anthony Watts
Joseph D'Aleo	Dr. Horst Lüdecke, PhD
George H. Taylor, Ms.	Sonja B-Christiansen, PhD
Arun Ahluwalia, PhD	Helmut Alt, Dr.
August Auer, MS	Robert C. Balling Jr., PhD
Jack Barrett, PhD	Bernard Beauzamy, PhD
David Bellamy, PhD	Ian Clark
Reid A. Bryson, PhD	Paul Copper, PhD
Alan Carlin, PhD EPA	Piers Corbyn, M.Sc.
Vincent Courtillot. PhD	David Deming, PhD
David Douglass, PhD	Reynald Du Berger, Ms.
Geoffrey Duffy, PhD	Hugh W. Ellsaesser, PhD
Robert H. Essenhigh, PhD	Christopher Essex, PhD
Samuele Furfari, PhD	Goklany, PhD
Gerhard Gerlich, Ph. D.	Lee C. Gerhard, PhD
Stewart Franks, PhD	Craig D. Idso, PhD
Stanley B. Goldenberg, MS	Sherwood B. Idso, PhD
Keith D. Hage, PhD	Keith E. Idso, PhD
Roger Helmer, MA	Richard A. Keen, PhD
Kiminori Itoh, PhD	Michael Joseph Kelly, PhD
Claes Göran L. Johnson, PhD	Steve Koonin, PhD
Wibjorn Karlen, PhD	Henri A. Masson, PhD
George Kukla, RNDr.	James E. McGrath, PhD
Gerrit van der Lingen, PhD	Fred Michel, PhD
Asmunn Moene, PhD	Tad Murty, PhD
Alan Moghissi, PhD	Gernot Patzelt, PhD
Nils-Axel Mörner, PhD	Al Pekarek, PhD
Harry N.A. Priem, PhD	Joanne Simpson, PhD
Arthur Rörsch, PhD	L. Graham Smith, PhD
Tom Victor Segalstad, PhD	Brian G. Valentine, PhD
Jan Veizer, PhD	David Whitehouse, PhD
James A. Wanliss, PhD	Antonino Zichichi, PhD
Werner Weber, PhD	Fritz Vahrenholt, PhD
Chris Landsea, PhD	Patrick Moore, PhD

The Right Climate Stuff Research Team

Harold H. Doiron, PhD

W. Cunningham, MS, Apollo 7

H. Schmitt, PhD Apollo 17

37 other former NASA employees

Notes

1. Remarks to the Commonwealth Club, https://bit.ly/2YexUVV

2. Roosevelt Institute, Rural Electrification Administration, https://bit.ly/2UWKKGM

3. Washington Post, April 20, 2020.

4. Bansky quotation, https://bit.ly/3grKB8j

5. MyClimate Website, https://bit.ly/2LRiGjP

6. Donn Dears. *Nothing to Fear.* Critical Thinking Press: Jan. 1, 2015.

7. Donn Dears. *Energy: The Source of Prosperity.* Donn Dears LLC: July 10, 2019. Chapter 8-2 covers this in more detail.

8. US Energy Information Assoc. (EIA), "Levelized Cost and Levelized Avoided Cost of New Generation Resources in the Annual Energy Outlook 2020," http://bit.ly/3b9VurR

9. Renewable energy credit, Wikipedia, Last edited on 26 March 2019, https://bit.ly/2AMx1fc

10. Abraham Lincoln quote, Goodreads, https://bit.ly/2UmpM4H

11. The transmission grid that the ERCOT independent system operator administers is located solely within the state of Texas and is not synchronously interconnected to the rest of the United States. The transmission of electric energy occurring wholly within ERCOT is not subject to the Commission's jurisdiction under sections 203, 205, or 206 of the Federal Power Act.

12. "The North American Electric Reliability Corporation (NERC) is a not-for-profit international regulatory authority whose mission is to assure the effective and efficient reduction of risks to the reliability and security of the grid. NERC develops and enforces Reliability Standards; annually assesses seasonal and long-term reliability; monitors the bulk power system through system awareness; and educates, trains, and certifies industry personnel. NERC's area of responsibility spans the continental United States, Canada, and the northern portion of Baja California, Mexico. NERC is the electric

reliability organization (ERO) for North America, subject to oversight by the Federal Energy Regulatory Commission (FERC) and governmental authorities in Canada. NERC's jurisdiction includes users, owners, and operators of the bulk power system, which serves more than 334 million people." https://bit.ly/2AiKhZ9

13. FERC Order 2000, https://bit.ly/2BNvPsz

14. California Assembly Bill 1890, "Electric Utility Restructuring: Maintaining Bulk Power System Reliability," Every CRS Report, 36, CRS-16, Service Obligations, February 1, 2005, https://bit.ly/2MD3pUl

15. Rick Glick and Mathew Christiansen, "FERC and Climate Change," The Energy Bar Assoc, 2019, https://bit.ly/2YaqR0q

16. The Renewable Electricity Production Tax Credit: In Brief, Congressional Research Service, https://bit.ly/2A5vvnW

17. Coal-fired power plants can't be built because they cannot meet current EPA regulations requiring that they emit no more than 1,400 pounds of CO_2 per MW-hour. The cost of current nuclear power plant designs are prohibitive and the Nuclear Regulatory Commission has not yet approved the design of any small modular reactor.

18. Voltages must be kept within certain limits to prevent damage to equipment. The formula $P=EICos\ \theta$, shows the relationship between voltage and power.

19. The Structure of Capacity Auctions, Penn State, College of Mineral and Earth Sciences, Course 9.2.2.

20. PJM, Wind Effective Load Carrying Capability (ELCC) Analysis, 9/13/2018, https://bit.ly/2UyDK3P

21. ibid.

22. 162 FERC ¶ 61,205, Docket No. Docket No. ER18-619-000, Order on Tariff Filing. (Issued March 9, 2018), https://bit.ly/2XMjgGy

23. From ISO-NE cold weather operations report, December 24, 2017, to January 8, 2018, http://bit.ly/2q7Eldb

24. ibid.

25. ibid.

26. ibid.

27. Glick and Christiansen, "FERC and Climate Change," The Energy Bar Assoc, 2019, https://bit.ly/2YaqR0q

28. Jeffrey Tomich, "Turbine shutdowns in polar vortex stoke Midwest debate." Energywire, Wednesday, February 27, 2019, https://bit.ly/3h6aEC4

29. FERC Order No. 2000, https://bit.ly/2BNvPsz

30. ibid.

31. ISO-New England, "Operational Fuel-Security Analysis," Jan. 17, 2018, http://bit.ly/2EDtQDx

32. Meredith Angwin, author of *Campaigning for Clean Air* and yet to be released *Shorting the Grid*, https://bit.ly/2AN8E14

33. Ethan Howland, "Federal Energy Regulatory Commission Keeps Even Partisan Split," Roll Call, Posted October 25, 2018, at 3:25pm, https://bit.ly/3f7sUJF

34. "FERC deadlock could continue past end of LaFleur's term," Iulia Gheorghiu@IMGheorghiu, May 7, 2019, https://bit.ly/2APDTIQ

35. Glick and Christiansen, "FERC and Climate Change," The Energy Bar Assoc, 2019, https://bit.ly/2YaqR0q

36. Jeremy Dillon and Rod Kuckro, "FERC commissioners grilled on climate," E&E News Published: Thursday, June 13, 2019, https://bit.ly/37akqP9

37. FERC Order No. 2000, https://bit.ly/2BNvPsz

38. FERC, "About FERC," https://www.ferc.gov/about/about.asp

39. FERC Strategic Plan, https://bit.ly/2UUCXcz

40. Stu Bresler, "The value of markets," PJM Inside Lines. May 22, 2018, http://insidelines.pjm.com/the-value-of-markets/

41. ibid.

42. GE Power. "Breaking the power plant efficiency record, GE & EDF unveil a game-changer at Bouchain," https://invent.ge/37emlSK

43. PJM. "PJM Offers Ideas on Carbon Pricing," PJM Inside Lines, August 24, 2017. From *Advancing Zero Emissions Objectives Through PJM's Energy Markets: A Review of Carbon-Pricing Frameworks.*

44. NYISO. "Reliability and a Greener Grid," Power Trends 2019. The New York ISO annual *Grid and Market Report*, 2019, https://bit.ly/3f4JuK5

45. NYISO. PRESS RELEASE | NYISO Releases Power Trends 2019: Annual State of the Grid & Markets Report, May 2, 2019, https://bit.ly/3cJ9fxV

46. History Matters, Franklin D. Roosevelt's first inaugural speech, 1933, https://bit.ly/2TzFSHQ

47. Cosmic rays refer to studies by Henrik Svensmark and Nir Shaviv.

48. "Record of volcanic eruptions during the Holocene at Smithsonian Institution," https://s.si.edu/3gk5NNi

49. Jule G. Charney, "Carbon Dioxide and Climate: A Scientific Assessment," Report of an Ad Hoc Study Group on Carbon Dioxide and Climate, Woods Hole, MA. July 23-27, 1979, https://bit.ly/2XMHX5H

50. IPCC. "Climate Change 2014 Synthesis Report. Fifth Assessment Report" © Intergovernmental Panel on Climate Change, 2020, https://ar5-syr.ipcc.ch/topic_summary.php

51. Chris Mooney and Andrew Freedman, "We may avoid the very worst climate scenario. But the next-worst is still pretty awful," The Washington Post, January 30, 2020, https://wapo.st/3dMcmXl

52. Testimony of John R. Christy, Professor of Atmospheric Science, Alabama State Climatologist University of Alabama in Huntsville, U.S. House Committee on Science, Space & Technhology, 29 Mar 2017, https://bit.ly/2ASFm0M

53. ibid.

54. Curry, Judith A. "Climate Models for the layman," The Global Warming Policy Foundation, 2017, https://bit.ly/2zfS8pT

55. Curry, Judith A. "Plausible scenarios for climate change: 2020-2050," Climate Etc., Feb. 13, 2020, https://bit.ly/3dMM47n

56. ibid.

57. "William Happer Talks Climate Alarmism During COP25 in Madrid." Heartland.org., Dec. 6, 2019, https://bit.ly/2UpbIan. Graph from Dr. W. Happer's press briefing, Madrid, Spain, Dec. 2019, https://bit.ly/2MFcCLL

58. ibid.

59. "Old carbon reservoirs were not important in the deglacial methane budget" Science 21 Feb 2020: Vol. 367, Issue 6480, pp. 907-910, https://bit.ly/2MIvrxM. See also, University of Rochester, Methane study, https://bit.ly/340OueB

60. ibid.

61. Quotations by Christiana Figueres, the executive secretary of the United Nations Framework Convention on Climate Change (UN-FCCC). "Figueres: First time the world economy is transformed intentionally," UNRIC. 3 February 2015, https://bit.ly/30r54Va See also *Nothing to Fear,* page 24, for full quotations.

62. Naomi Klein. *This Changes Everything: Capitalism vs The Climate,* Simon & Schuster; Reprint edition (August 4, 2015), https://bit.ly/3h9Bc5x

63. Quote by Ottmar Edenhofer, IPCC, from "IPCC Official: 'Climate Policy Is Redistributing The World's Wealth,'" The Global Warming Policy Forum, Nov. 11, 2010, https://bit.ly/2UppBW5

64. K. Bithas and P. Kalimeris, "Revisiting the Energy-Development Link," Springer Briefs in Economics. Source: Bithas and Kalimeris, 2016, Figure 2.1, p. 8. 26, DOI 10.1007/978-3-319-20732-2_2

65. World Bank Policy on financing coal-fired power plants. https://bit.ly/2MUIpZe

66. McKinsey on Powering Africa, https://mck.co/2zAmQKw

67. Electricity consumption 2020 by Country, https://bit.ly/3ft3er3

68. Asanta Sana (Thank you Very Much!): Life Lessons from Kili, By Dr. Cheryl McAuley, page 45.

69. Lake Turkana wind farm in Kenya, https://bit.ly/2UJUfcV

70. Raymond L. Gifford and Matthew S. Larson, "State Actions in Organized Markets, States Strive to 'Fix' Markets and Retain Base Load Generation," page 14, http://bit.ly/2O1GzWO

71. Frank Wouters and Ad van Wijk "50% Hydrogen for Europe: a manifesto," Energy Post.eu. May 7, 2019, https://bit.ly/3f2MkPF

72. M. W. Melanina, O. Antonia, and M. Penev, "Blending Hydrogen into Natural Gas Pipeline Networks: A Review of Key Issues," National Renewable Energy Laboratory (NREL) Technhical Report , NREL/TP-5600-51995 March 2013, https://www.nrel.gov/docs/fy13osti/51995.pdf

73. Woulters and van Wijk, "50% Hydrogen for Europe."

74. "Natural gas pipeline profits, construction both up," Oil & Gas Journal, Sep. 5, 2016, https://bit.ly/2YjCHFG

75. Nathan Parker, "Using Natural Gas Transmission Pipeline Costs to Estimate Hydrogen Pipeline Costs," Institute of Transportation Studies, UC Davis, 2004, 2004_UCD-ITS-RR-04-35.pdf

76. World Steel Assoc. "Steel's Contribution to a low carbon future," World Steel Association. worldsteel.org, https://bit.ly/3bsd3mT

77. Rubenstein, Madeleine. "Emissions from the Cement Industry" May 9, 2012, State of the Planet Earth Institute, Columbia University. Columbia University, https://bit.ly/3bpmKCL

78. Sören Amelang, "Steelmaker Salzgitter expects policymakers to help fund massive decarbonisation project," Clean Energy Wire, 04 Apr 2019, 14:09, https://bit.ly/30riHDQ

79. Article on cement: Earth Institute, Columbia University, "Emissions from the Cement Industry," https://bit.ly/3bpmKCL

80. "Lazard's Levelized Cost of Energy (LCOE) analysis," Nov. 7, 2019, https://bit.ly/3cQZduG

81. Donn Dears, *Energy: The Source of Prosperity*. Donn Dears LLC: July 10, 2019, Chapter 9-3.

82. ibid.

83. Brian Lips. "Credit Multipliers in Renewable Portfolio Standards," prepared for the RPS Collaborative, The Clean Energy States Alliance. July 2018, http://bit.ly/2rt9nQg

84. Henry Hub Natural Gas Spot Price, USEIA, https://bit.ly/2XKZs67

85. Global Warming Petition Project, http://www.petitionproject.org

Index

Page **numbers in boldface** refer to main discussions; page *numbers in bold italics* refer to graphs, tables, and other illustrations.

Q

About the Author

Donn Dears

Donn Dears is a retired GE Company senior executive specializing in power generation, with extensive experience in Europe, the Mideast and Southeast Asia. He worked with government officials to obtain needed permissions to establish GE subsidiaries in fourteen countries to service large power generation and related electric apparatus.

Donn began his career at General Electric as a test engineer testing large steam turbines and generators used by utilities to generate electricity. He then spent three years on GE's prestigious Manufacturing Management Program in diverse businesses, including locomotives, DC motors, medium steam turbines, small jet engines, and naval ordnance businesses. This was followed by five years in manufacturing and marketing assignments at the Transformer Division.

He then led organizations servicing GE power generation and large equipment in the United States and then around the world. Donn was involved with work done at customer locations, such as steel mills, electric utilities, refineries, oil drilling and production facilities and open pit and underground mining operations. He subsequently led an engineering department supporting GE organizations and subsidiaries around the world.

Donn has traveled extensively, beginning with a year at sea while a cadet-midshipman at the United States Merchant Marine Academy where he visited countries in Asia, Europe, and South America while still nineteen. He continues to travel and has visited over sixty countries on business and for pleasure. Donn is a graduate of the U.S. Merchant Marine Academy and served on active duty in the U.S. Navy, first as damage control officer, then as engineering officer during the Korean war.

Donn blogs, writes articles, and speaks on the issues of climate change and the vital need for low cost, reliable electricity.